Sustainable Construction in the Era of the Fourth Industrial Revolution

This book provides readers with an understanding of various concepts, benefits, and practices that the adoption of Fourth Industrial Revolution (4IR) technologies can bring when working towards sustainable construction goals.

As digitalization continues to advance rapidly, the pressures on stakeholders in the architecture, engineering, construction, and operation (AECO) industry to revamp and restructure their activities and outputs become increasingly prevalent. This research book explains the importance of various digital tools and principles to achieve sustainable construction projects. It adopts various standards and concepts to highlight how 4IR technologies could assist and accelerate construction sustainability. It is the first book to link construction management with various digital tools to enhance construction projects' sustainability. It also provides an in-depth insight into the concept of sustainable construction 4.0 across both developing and developed countries for construction professionals, sustainability experts, researchers, educators, and other stakeholders.

The book can be adopted as a research guide, framework, and reference on sustainable construction, the concept of sustainable projects, digitalization in the construction industry, and the 4IR.

Ayodeji Emmanuel Oke is a Senior Lecturer in the Department of Quantity Surveying, Federal University of Technology Akure, Nigeria and a Research Fellow in the Department of Construction Management and Quantity Surveying, University of Johannesburg, South Africa. He has taught several modules in value management, quantity surveying, and sustainable construction management for some years. His specialization areas are in construction sustainability management, with emphasis on value management, quantity surveying, and construction in the digital era.

Clinton Ohis Aigbavboa is a Professor in the Department of Construction Management and Quantity Surveying and Director of cidb Centre of Excellence & Sustainable Human Settlement and Construction Research Centre, University of Johannesburg, South Africa. Before entering academia, he was involved as a quantity surveyor on several infrastructural projects in Nigeria and South Africa. He has published several research papers and more than ten research books in housing, construction and engineering management, and research methodology for construction students. He is the editor-in-chief of the *Journal of Construction Project Management and Innovation*.

Seyi Segun Stephen is a graduate of Quantity Surveying at the Federal University of Technology Akure, Nigeria. He is a student coordinator for the Civil Engineering Measurement course at the Federal University of Technology, Akure, Nigeria. He is also an online member of the Young Quantity Surveyors Forum to date and the student coordinator at the Ministry of Works Lands and Housing, Akure, Nigeria.

Wellington Didibhuku Thwala is a Professor in the Department of Construction Management and Quantity Surveying and Chair of SARChI in Sustainable Human Settlement and Construction Research Centre, University of Johannesburg, South Africa. He has extensive industry experience with a research focus on sustainable construction, leadership and project management. He is the editor-in-chief of the *Journal of Construction Project Management and Innovation*. He also serves as an editorial board member to various reputable international journals.

Routledge Research Collections for Construction in Developing Countries

Series Editors: Clinton Aigbavboa, Wellington Thwala, Chimay Anumba, David Edwards

Sustainable Construction in the Era of the Fourth Industrial Revolution

Ayodeji E. Oke, Clinton O. Aigbavboa, Seyi S. Stephen and Wellington D. Thwala

Routledge
Taylor & Francis Group

LONDON AND NEW YORK

First published 2022
by Routledge
2 Park Square, Milton Park, Abingdon, Oxon OX14 4RN

and by Routledge
605 Third Avenue, New York, NY 10158

Routledge is an imprint of the Taylor & Francis Group, an informa business

© 2022 Ayodeji E. Oke, Clinton O. Aigbavboa, Seyi S. Stephen and Wellington D. Thwala

British Library Cataloguing-in-Publication Data
A catalogue record for this book is available from the British Library

Library of Congress Cataloging-in-Publication Data
Names: Oke, Ayodeji E., author. | Aigbavboa, Clinton, author. | Segun, Stephen, author. | Thwala, Wellington, author.
Title: Sustainable construction in the era of the fourth industrial revolution / Ayodeji E. Oke, Clinton O. Aigbavboa, Stephen S. Segun and Wellington D. Thwala.
Description: Abingdon, Oxon ; New York, NY : Routledge, 2022. | Series: Routledge research collections for construction in developing countries | Includes bibliographical references and index.
Subjects: LCSH: Sustainable construction. | Industry 4.0.
Classification: LCC TH880 .O43 2022 (print) | LCC TH880 (ebook) | DDC 624.028/6--dc23
LC record available at https://lccn.loc.gov/2021011313
LC ebook record available at https://lccn.loc.gov/2021011314

ISBN: 978-1-032-01215-5 (hbk)
ISBN: 978-1-032-01755-6 (pbk)
ISBN: 978-1-003-17984-9 (ebk)

DOI: 10.1201/9781003179849

Typeset in Goudy Oldstyle Std
by KnowledgeWorks Global Ltd.

To God

Who Made All Things Beautiful

Contents

Preface

The roles played by the competent professionals involved in construction are paramount to the achievement of project objectives at the completion of each project. This has motivated stakeholders concerned with construction projects to keep learning and developing their skills to better offer services that will enhance the performance of projects. To strike a balance among various objectives of infrastructural development, construction professionals have resorted to embracing the use of information technology and digital tools to improve every aspect of construction from reconnaissance and inception to completion, usage, maintenance, and final stage. The digital tools are basic elements of the Fourth Industrial Revolution (4IR) such as gamification, machine learning, artificial intelligence, and augmented reality, among others. Adopting these in the construction industry is termed construction 4.0.

There have been many sustainable practices that have supported the effective production of infrastructure. One such sustainable practice is information technology that has been incorporated into constructional practices. Sustainable construction provides a sustained framework that is centred around an environment that is utilized and maximized to the highest possibilities. It encompasses the introduction of various measures to make preventive or corrective applications to infrastructural development for the infrastructure to be able to provide the service it is due for within the expected life cycle. Sustainable construction processes have been applied in construction projects across the world, in both developed and developing countries, to improve the economic situation through the set out means. This is however not the case in some under-developed countries where there is little or no such practice; these countries work more towards construction that satisfies the immediate wants and needs of the clients and meets the supply of their basic infrastructure needs.

In order to raise the awareness of the general populace regarding its further implementation along with the benefits that accrue with it, it is necessary to the various drivers, concepts, and steps by which its nearly total implementation would benefit the construction industry significantly. This book therefore provides the readers with an understanding of numerous concepts, benefits, and practices that the adoption of the 4IR technologies would bring when working towards a sustainable construction. The adoption of 4IR principles for sustainable construction is termed sustainable construction 4.0.

As digitalization continues to advance progressively, the pressures on stakeholders in the architecture, engineering, construction, and operation (AECO) industry to revamp and restructure their activities and deliveries become increasingly urgent. This research book on sustainable construction in the era of the 4IR explains the importance of various digital tools, also known as 4IR principles or ICT tools in the achievement of sustainable construction, hence the term 'sustainable construction 4.0'. The expected readers of this book include built environment scholars; government agencies (public clients) such as parastatals, ministries, and other arms of the government that are concerned with the infrastructure and other related developmental projects; corporate agencies involved in planning, execution and infrastructural management; individual clients who desire to have a sustainable project; owners of construction projects; policy makers who are concerned with improving the quality and performance of construction projects; construction professionals charged with the responsibilities of monitoring and developing construction works; bodies and boards involved in monitoring and regulating the professionals; building contractors in various categories of project execution in building, civil, and industrial engineering areas; and financiers of construction projects, including banks, insurance companies, bond companies, and loan firms, amongst others.

The book adopts various standards and concepts in bringing out the ways by which the 4IR technologies could assist in achieving construction sustainability. These concepts are divided into parts/sections with various chapters that are all expressed in terms of the 4IR fusion in the achievement of sustainable construction. Each of the chapters starts with an introduction which introduces the knowledge relating to the concept to be discussed and ends with a conclusion that summarizes the major highlights of the concepts and their potential in support of sustainable construction. As this is a research book, references are provided at the end of each chapter for additional information about the subject. An index of important and key terms is also provided to enable a swift check of an identified area of study.

The expected readers of this book include built environment scholars; government agencies (public clients) such as public firms, ministries, and other arms of the government that are concerned with infrastructure and other related developmental projects; corporate agencies involved in planning, execution, and infrastructural management; individual clients who desire to have a sustainable project; owners of construction projects; policy makers who are concerned with improving the quality and performance of construction projects; construction professionals charged with the responsibility of monitoring and developing construction works; bodies and boards involved in monitoring and regulating the professionals; building contractors in various categories of project execution in building, civil, and industrial engineering areas; and financiers of construction projects including banks, insurance companies, bond companies, and loan firms, amongst others.

The book can be adopted as research guide, framework, and note on material topics relating to sustainable construction, the concept of sustainable projects, project performance indices, and sustainable developments in the construction industry. We hope that all readers of this book will find it educating, interesting, and impacting in shaping their knowledge in understanding the 4IR technologies for sustainable construction.

Ayodeji E. Oke
Clinton O. Aigbavboa
Seyi S. Stephen
Wellington D. Thwala

Part I

Background information
of the book

1 General introduction

Introduction

Construction has moved from the stage where buildings are constructed with any material available to cater for an immediate need. It has moved to a better state of erecting structures that are not only smart but also sustainable enough to prevent cost overrun and material wastage. With several policies aiming to improve the economic conditions of a particular region, working in the way of a sustainable construction has been identified as propelling and fast tracking the policies in proposal and implementation towards a more stable economy. In the past, construction practices were saddled with the use of crude materials which, in turn, made construction expensive. The inadequacy in the cost management and that of effective all-round management of construction practice has driven the construction professionals to consider ways of implementing information technologies into construction. The implementation has not been easily accepted as many professionals believe that the disadvantages involved in its full implementation outweigh the services it is going to render in construction. This gave rise to some construction firms' sticking to the tried and tested ways of construction with little interest in the adoption of information technology-related concepts.

The various measures used mostly in traditional construction are dependent on human efforts and understanding that are always prone to errors as a result of fatigue, overloading, monotonous practices, and other factors that are concerned with the limited output experienced when human labour is the main source of project execution. Since the advancement in Internet facilities, clients have developed an interest in having projects delivered within the shortest possible time as they have access to information regarding construction projects delivered within days in some advanced countries. This has led to environmentalists adopting the use of machine-related technologies to accelerate project execution while simultaneously saving cost, energy, and time. Moreover, buildings are now designed to meet environmental standards in terms of aesthetics, green option, accessibility, flexibility, land use, duration of whole life cycle, and life costing. In order to meet up with this reality, the human service is becoming a liability and engaging drones, robotics, artificial intelligence, and big data, amongst others, are

DOI: 10.1201/9781003179849-1

becoming necessary with every construction demand. There are other practices that are employed to get the best out of construction; however, one that will last for a long time is the adoption of information technologies as it cuts across every bit of information needed to execute and manage a particular project. This book focuses on exploring the adoption and usage of information technology for the actualisation of sustainable infrastructural development in the construction industry.

The book is divided into 2 parts and 16 chapters for ease of use and guidance to the reader. The first part explains the background information of the book to introduce the general idea behind the topic. The second part contains the chapters that are further divided into subchapters as they relate to each identified chapter in information technology for sustainable construction. Chapter 1 explains measures as roles, concepts, applications, benefits, and challenges of artificial intelligence in sustainable construction. The next chapter focuses on the working principles of augmented reality towards a sustainable construction. In discussing big data and its influence on sustainable development, a detailed application of big data in infrastructure and in engineering analysis to bring out working principles is addressed. Chapter 4 further introduced building information modelling as it has been used for some time now. Its efficiency in the pre-design, design, and construction stages coupled with benefits, challenges, and concepts comprise this chapter. Biomimicry comprises Chapter 5, along with its designs and considerations for improving building sustainability.

In discussing blockchain as another information technology concept, the implementation of other technologies, discussed in Chapters 1–5, is further buttressed with the highlighted benefits of blockchain and how it could impact the construction industry. The pillars of sustainability are driven to meet a specific need and cryptocurrency is not excluded from this. How cryptocurrency serves as a medium of exchange and means of payment to making transactions faster and secured through identified channel is detailed. Chapter 8 gives much detailed measures, examples and steps towards a construction industry that is safe from threats. Chapter 9 deals with the drone in construction. Its application ranges from site survey, safety management, inspection, quality management, time management and site management to demolition management. Apart from its application, its impact is also discussed to give a better understanding of the role it plays towards a sustainable environment. Chapter 10 discusses gamification in exploring options that may not be totally visible when considering sustainable construction in the first phase of the project. Drivers and challenges of applying gamification, advantages of gamification, gamification for construction sustainability, the relationship between gamification and sustainability, and gamifying construction are some of the features associated with gamification in the construction industry.

To enhance and raise awareness of the value associated with sustainable construction, Chapter 11 is designed to bring into notice the Internet of Things as well as related models for improved construction. Chapter 12 discusses machine learning benefits and limitations in the construction industry, along with other highlighted factors. In getting the best outcome possible from the use of advanced and heightened materials, Chapter 13 discusses nanotechnology to solidify and improve strength in the construction industry. Chapter 14 details the ease and diversity brought into construction through the application of robotics to perform various functions in achieving sustainable construction. The last chapter of this book discusses the prospects of adopting virtual reality for construction works. Benefits of utilising virtual reality within the construction industry include viability and feasibility appraisal, modifying the design at any stage of the construction, safety and construction practicalities, improved project planning, enhanced collaboration, and other indirect benefits.

Sustainable construction

In understanding sustainability in the construction practices and methods attached to it, the basic principle of constantly improving the quality of construction should always top the agenda list. However, depending on an individual's understanding of sustainability, it is said to be focused on the goal of saving aspects of construction that has received little consideration in the past. In all the chapters of this book, the different/various technologies used in the construction industry are aimed at bringing to attention the advanced benefits of involving these simple but effective means of construction. The objective is to enhance, solidify, improve, and more essentially bring into focus a new understanding of what management is all about. In respect of the sustainable technology application in construction and other industries, authors such as Lee, Yoon, and Kim, (2007); Zhang, Platten and Shen, (2011); Roufechaei, Bakar and Tabassi, (2014); Chen, Yang and Lu, (2015); Koebel et al. (2015); and Ahmad, Thaheem and Anwar, (2016) have discussed the application of sustainable technology in quality environment enhancement and resources and materials efficiency amongst other benefits.

The concept of sustainable construction has grown over the years as many see the need for it to be taken seriously along with its integration that the industry has witnessed since its adoption. The enhanced visibility of sustainable construction has further raised another question regarding what the correlation is between sustainability and management. While most construction stakeholders are working towards a manageable cost construction, sustainability raises the issue of the consistent value of cost to be obtained in the long run or the estimated duration in order to get the best maintenance for their money. Sustainable construction exceeds the scope of ordinary cost and management; it expresses construction in ways it relates to everything feasible right from the onset to the estimated duration of the construction (whole life concept). The key focus in achieving sustainable construction is bringing together all the construction professionals in a common understanding of what sustainability comprises. These professionals can then conceptualise their understanding of sustainability and how it can be implemented effectively in their respective firms.

Need for information and communication technologies in construction

In trying to resolve some of the constraints in construction, the adoption of information technology has been significant. Studies by Hassanain, Froese and Vanier (2000); Peansupap (2004); Gopalakrishnan and Brindha (2017); and Mohammad et al. (2017) have identified time constraints, the nature of work in terms of complexity, and breaking down of operations into units as having led to large, medium, and small organisations' adopting the implementation of information technologies for most of their organisation's activities. It is clear that the implementation of these technologies is centred towards the aiding adequate sharing of information from the lowest level of employees to those in top positions. In cases where there is proper dissemination of information, the operations of the project run smoothly since everyone knows what is expected of them from the onset of the project. Moreover, in sustaining construction, job of construction professionals and the environment at large, information and communication enhanced technologies are paramount in making the environment more habitable for everyone. The essentiality of information and communication technologies (ICT) cannot be underestimated as it covers for monument increment in impacts and activities in stages of construction from planning to site operations. It further expands the knowledge of the industry in aspects such as the following:

- Project management
- Contract resolutions
- Legal matters
- Material and labour management
- Safety protocols
- Supply chain management
- Monitoring and performance measurement
- Cost management
- Differentiation and building product customization
- Quality projects
- Whole life cycle

Objective of the book

There are many materials that are in existence on sustainable construction, information technology, and information communication technology in construction and other related subjects (Carter, Hassan & Merz, 2001; El-Ghandour & Al-Hussein, 2004; Andrews, Rankin & Waugh, 2006; Lee et al., 2007; Jacobsson & Linderoth, 2010; Zhang et al., 2011; Sextos, 2014; Ahmad et al., 2016; Chen et al., 2015; Martínez-Rojas, Marin & Vila, 2015). These materials include journals, textbooks, conference papers, personal publications, research books, and book chapters that express the information regarding each identified subject area.

This research book on the application and enhancement of information technology as a means to sustaining construction practices does not only serve as a tool, but it also expatiates the aspect of construction that could be channelled towards a better environment and economy for the stakeholders involved and the entire population. The book explains sustainable construction as that which is germane to an all-round package in delivering, relieving, ameliorating, and as well as bringing stability to the construction discipline and to projects at the same time. This book promotes the knowledge of applicable measures to working towards an environment and construction that is sustainable. The book also provides additional input to researchers that are concerned with sustainable construction, employers of labour in the industry, construction professionals, concerned persons, and students. It improves the understanding of the concept of sustainable construction, the benefits of engagement, and other advantages related to making the environment safer of putting resources that are in line to making the environment safe, sound, beneficial, and environmentally sensitive to the direction positive than what we have now.

Summary

This chapter introduces general information about the reason for the conception of the book with relevance placed on expatiating issues relating to sustainable construction 4.0. In addition, the need for more incorporation of information technology into construction was emphasized if construction stakeholders wish to bring their projects in line with sustainable practices. The need for incorporating fourth industrial revolution (4IR) principles into construction underpins the researchers' detailed explanation of its benefits. Moreover, the possible consequences of the industry's continuation with the present approach of handling construction projects were spelt out. The objective of this book is to promote the adoption of information technology by construction professionals regarding the principles that are related to construction projects. This would lead to efficient projects that all could be proud of. This objective is also explained in each of the chapters by detailing relevant information about the various technologies used in the construction industry.

References

Ahmad, T., Thaheem, M. J. & Anwar, A. (2016). Developing a green-building design approach by selective use of systems and techniques. *Architectural Engineering and Design Management*, 2(1), 29–50.

Andrews, A., Rankin, J. H. & Waugh, L. M. (2006). A framework to identify opportunities for ICT support when implementing sustainable design standards. *Journal of Information Technology in Construction*, 11, 17–33.

Carter, C., Hassan, T. & Merz, M. (2001). The eLegal project: Specifying legal terms of contract in ICT environment. *International Journal of Information Technology in Construction*, 6, 136.

Chen, X., Yang, H. & Lu, L. (2015). A comprehensive review on passive design approaches in green building rating tools. *Renewable and Sustainable Energy Reviews*, 50, 1425–1436.

El-Ghandour, W. & Al-Hussein, M. (2004). Survey of information technology applications in construction. *Construction Innovation*, 4(2), 83–98.

Gopalakrishnan, G. & Brindha, G. (2017). A study on maternity benefit and its effectiveness in construction industry. *International Journal of Civil Engineering and Technology*, 8(10), 130–136.

Hassanain, M. A., Froese, T. M. & Vanier, D. J. (2000). IFC-based data model for integrated maintenance management. *Computing in Civil and Building Engineering*, 2000, 796–803.

Jacobsson, M. & Linderoth, H. (2010). The influence of contextual elements, actors' frames of reference, and technology on the adoption and use of ICT in construction projects: A Swedish case study. *Construction Management and Economics*, 28(1), 13–23.

Koebel, C. T., McCoy, A. P., Sanderford, A. R., Franck, C. T. & Keefe, M. J. (2015). Diffusion of green building technologies in new housing construction. *Energy and Buildings*, 97, 175–185.

Lee, S. K., Yoon, Y. J. & Kim, J. W. (2007). A study on making a long-term improvement in the national energy efficiency and GHG control plans by the AHP approach. *Energy Policy*, 35(5), 2862–2868.

Martínez-Rojas, M., Marin, N. & Vila, M. A. (2015). The role of information technologies to address data handling in construction project management. *Journal of Computing in Civil Engineering*, 30(4), 04015064.

Mohammad, W., Shumank, D., Raj, B. D. & Mohd, B. K. (2017). A study of project success and procurement frameworks in Indian construction industry. *International Journal of Civil Engineering and Technology*, 8(3), 167–174.

Peansupap, V. (2004). An exploratory approach to the diffusion of ICT in a project environment. PhD, RMIT University, School of Property, Melbourne.

Roufechaei, K. M., Bakar, A. H. A. & Tabassi, A. A. (2014). Energy-efficient design for sustainable housing development. *Journal of Cleaner Production*, 65, 380–388.

Sextos, A. (2014). ICT applications for new generation seismic design, construction and assessment of bridges. *Structural Engineering International*, 24(2), 173–183.

Zhang, X., Platten, A. & Shen, L. (2011). Green property development practice in China: Costs and barriers. *Building and Environment*, 46(11), 2153–2160.

Part II

Sustainable construction 4.0

2 Artificial intelligence for sustainable construction

Introduction

Technology is as old as man, and the invention of the wheel dates back to over four millennia, while the revolution of transportation was associated in relation to the archaic pyramids in Egypt. The twenty-first century has witnessed unprecedented technological advancement with artificial intelligence (AI) setting the pace in revolutionising the construction industry and those sectors associated with it. This has led to world economic growth. However, AI has its own challenges and benefits though. Some hundreds of years ago, modern plant was not available and the construction industry depended on manual labour (Newton, 2018). AI consists of electronic devices fused together with software-driven systems that relate to the environments they occupy in construction and act to elicit optimal performance within a given set of conditions (Adio-Moses & Asaolu, 2016). Newton (2018) defined AI as the ability of computer systems to perform tasks that require normal human intelligence to perform. According to Newton (2018), AI extends beyond the use of automation in construction and includes the usage of off-site modular construction. This creates economies of scale in design and manufacture, reduced labour costs and wastage as well as saving time, and improving quality; this is exemplified in the prefabricated London's Leaden Hall Building that was completed in 2014. In addition, 3D printing was used in the construction of a five-storey structure in Suzhou, China, while drones are useful in surveying, monitoring of construction site activities, and updating of safety records or material ordering. The aim of using AI is centred on the perception and capacity to move and manipulate things coupled with the natural language processing (communication) that relates to basic reasoning, planning, and conceptual understanding. The following are the approaches currently used:

1 Computational intelligence
2 Traditional symbolic
3 Statistical methods AI

There are many tools used today in AI, which encompasses the following:

1 Methods based on probability and logic and on probability
2 Neural networks
3 Versions of search and mathematical optimization

DOI: 10.1201/9781003179849-2

The AI field comprises many disciplines in which the following sciences and professions converge:

- Artificial psychology
- Computer sciences
- Linguistics
- Mathematics
- Psychology
- Philosophy and neuroscience

AI is also profoundly useful in form of intelligence sensors such as the UtterBerry sensor which provides real-time information pertaining to inaccessible parts of structure and gives data on structural defects. It influences engineering design as well as assisting in planned maintenance and scheduling, monitoring risks associated with costs, and delay. Building information modelling (BIM) enables designs to be done quickly. The models can be used for cheaper and more accurate facilities maintenance. It generates data for e-measurement and e-costing that assist in saving time and overhead costs of staffing.

Pitt et al. (2009) mentioned sustainable development to be one of the cogent issues to be reckoned with as it has reached its peak in the twenty-first century. This has attracted attention to look for a solution to problems threatening the planet in terms of resources depletion, global warming, irresponsible urbanisation, deforestation, as well as other issues. It has been observed that the total annual emissions of carbon dioxide have increased from 1.5 million tons in the 1950s to almost 6 billion tons in the 1990s. The study by Nikolaou et al. (2004) and Pitt (2009) gave insight into the intensiveness of the resources used in operations and constructions as they are liable to an output of great quantities of emissions in the process of construction, during the duration of use, and also at the stage of alteration or total demolition. Since the awareness of environmental performance increases, contemporary buildings should be planned and constructed towards a sustainable objective. Smart buildings are currently taking over and the futuristic advantages in terms of growth are feasible. It is, however, essential to consider and assess its approach and operations towards a sustainable future.

Concept definitions

Artificial intelligence

AI is defined in various understandings by various researchers. The AI definition by Luger and Stubblefield (1993) is expressed as the branch of computer science that is concerned with the automation of intelligent behaviour. Schalkoff (1990) explained AI as a field of study that seeks to explain and emulate intelligent behaviour in terms of computational processes while Kurzweil (1990) contends that AI is the art of creating machines that perform functions that require intelligence when performed by people.

AI has been demonstrated as a machine that has a perception of its immediate environment and that acts to achieve its maximum goal success. AI is a frequently used description for machines or computer-aided systems or devices that function like humans in resolving problems (Russell & Norvig, 2009).

Application of artificial intelligence in building and construction

The application of AI in the building and built industry is comparatively small in relation to other sectors such as banking, finance, as well as healthcare. Figure 2.1 explains the areas of application of AI in building and construction. AI has found application in the following areas of construction (Xmaterials, 2018).

Planning and design

AI planning and architectural design programmes are an effective tool for designing, planning, constructing, and managing construction projects. An AI sensor is used in building site surveys to obtain data for 3D blueprint maps, printing, and construction designs swifter than in the case of a human survey team.

Project management and administration

Construction tasks control and project management can be done with an AI system into which the sick days of an employee can be imputed and the system

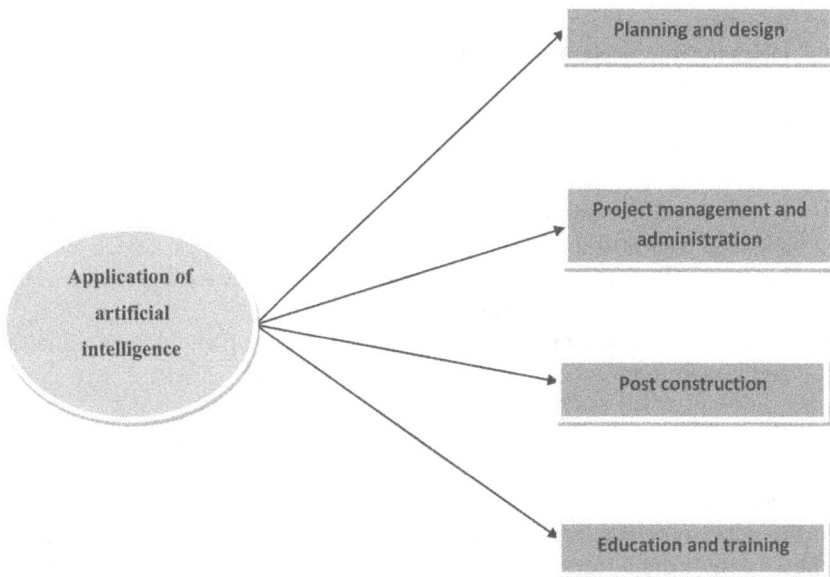

Figure 2.1 Applications of artificial intelligence in construction.

will reallocate his or her task to other employees for continuous update and improvement.

Post-construction

Architectural intelligence can be installed in completed building structures to control the lighting, temperature, and audio and visual installations through voice command. This is already in use in Amazon's Alexa system for hotels that is customisable to contain important information about guests such as check-out time and requests for room service.

Education and training

AI can also be used in image recognition software to analyse work-site unsafe behaviour of workers and the data obtained can be used by construction firms in improving on their training of staff.

Benefits of applying artificial intelligence to construction

According to Rao (2019), the benefits accruing from the application of AI to construction are outlined below:

- Prevents project cost over-runs by expediting project delivery time,
- Produces optimal solutions in building designs and helping to prevent rework,
- Assists in risk mitigation by helping in monitoring and prioritisation of project risks,
- Helps in projects planning and work scheduling by optimising and correcting the best project path,
- Boosts productivity and promotes efficiency on project sites,
- Ensures construction safety through image trackers that monitor areas of potential safety threats,
- Aids better distribution of equipment and labour across jobs by locating and addressing areas of shortage,
- Enhances efficient construction of off-site building components that are later fitted together by workers on site, and
- Generates large volume of data that construction stakeholders can analyse and benefit from.

Full acceptance of the application of latest technologies by building professionals is considered to be a daunting task as some prefer the traditional ways of construction. However, the emergence of AI and other modern systems is making construction more efficient, cost effective, and swifter in the delivery processes. The success of the introduction of AI in other industries has paved the way for more acceptance and usage in the built industry.

Prevent cost overruns in industry

The merit of using AI also extends to forecasting immediate or future cost overruns based on the available factors and estimates. These factors cut across the size of the project, type of contract, the professional status of the project managers, and other professional activities involved in the project. It makes use of historical data that includes commencement and completion dates using predictive models to draw up appropriate schedules and duration for future projects.

AI for better design of buildings through generative design

Apart from preventing or predicting cost overrun in projects, AI takes into consideration the pre- and post-activities, namely the planning and designing of the building to be constructed. It also considers the 3D models in electrical, architectural, mechanical, engineering, and plumbing (MEP) plans as well as the roles and functions of the identified teams in the project.

Risk mitigation

Risk is paramount in building a construction project. Risks are mostly directed towards the quality of workmanship and materials, the safety of the workers, and the duration agreed in the contract. From this, an inference is drawn that the larger a construction project, the relatively larger is the risk attached to it and vice versa. The involvement of AI reduces the risk, identified or unidentified, to the barest minimum by considering all-round operations that are concerned with the highlighted project.

Project planning

In identifying small or less significant problems, AI helps before they escalate into something serious. The management team in turn resorts to sorting out such problems before a situation is reached whereby corrective measures have to be applied. It is proposed that a later algorithm to be embedded in AI will be termed 'reinforcement learning'.

Sustainable construction

The Lafargeholcim Foundation (2019) stated that sustainable construction is geared towards requirements for housing, environment, and infrastructure that must provide the shelter, workspace, and the required services without impacting negatively on the ecological, social, and economic environment.

Sustainable construction is further defined as the search for buildings with energy efficiency, built with recycled materials, possessing a healthy and comfortable internal environment, safe with reduced construction, maintenance and life-cycle costs while maintaining the correlation between the ecological, social, and economic factors that are likely to be impinged on by the building (Van Dessel & Putzeys, 2007).

The US Environmental Protection Agency (EPA) defines it as the creation of structures by employing environmentally friendly processes and efficient resources from the beginning of the project to completion and throughout the entire life-cycle of a building (EPA, 2016). In order to inculcate a practice into a system, there must be a plan to integrate such a practice into the system's policy within or outside of a fiscal year. The same applies to sustainability; for it to be used properly, companies or firms should make executive plans to integrate their plans towards it. Effective sustainability practices highlight the fusing of sustainable plans and the implementation of such plans in order to obtain a sound project at the end of the day. The strategy to be employed by an organization should be well understood by the top management staff as well as those at the operational level. This relates better to the scheduled plans that have been formulated from the onset to completion.

Sustainable development is an approach that links the crucial factors of environmental, societal, and economic concerns that form part of a sustainable construction approach that seeks to meet the UNDP's sustainable development goals. Apart from performance, quality, and cost, there are six set principles for sustainable development that are conservation resources by minimizing their consumption, maximizing the resurge of resources, protecting the natural environment, creating a functional and non-toxic environment, using renewable resources, and creating a qualitative built environment.

Benefits of sustainable construction

Benefits of sustainable construction include the following:

i   Lower construction cost resulting from the application of sustainable construction technology such as AI,
ii   Better building designs due to cooperation between designers and constructors of construction projects,
iii   Environmental protection due to contractors' willingness to use sustainable and local materials and their desire to reduce waste and water pollution,
iv   Market expansion as witnessed in the rise of the global green building market to more than 260 billion dollars, and
v   Delivery of sustainable projects.

Artificial intelligence for sustainable construction

Christian Rauch (2018) pointed out that the constant evolvement of AI is among the most significant historical events in the life of man. Man is today almost at the peak of the industrial revolution that results from transformations caused by new technologies. Among these is AI that can play a notable role in sustainable development. AI uses key domain features in the area of natural language such as pattern recognition, machine translation, or computer vision with image recognition, voice recognition, classification, and data analysis.

According to Jennifer (2021), it was pointed out the United Nations AI Summit in Geneva in 2017 that AI has the potential to hasten a dignified life's progress and there should therefore be a focus on sustainable development in order to conserve natural environmental resources. How can AI be used for sustainable construction? The ideas are varied and many and are discussed in subsequent sections.

Use of satellite data

Christian Rauch (2018) identified one exciting area as the use of satellite data. Christian stated that the United States NASA start-up laboratories house over 200 mini-satellites which orbit the earth to obtain data which companies can access. In addition, they can order some custom features to obtain real-time monitoring. Raghav Bharadwaj (2019) indicated four areas of application of AI for sustainable construction as discussed in the following sections.

Construction planning and design

BIM uses a generative approach to mechanical, electrical, and plumbing (GenMEP) to execute planning and design. The effective planning of a project, designing, management, and construction are made more visible by what BIM offers. These benefits of using the BIM are made possible because of its usage of a 3D model implemented process. The Autodesk Revit is a BIM software which enables a 3D design of buildings and components, linking them up with time-related information.

Building System Planning with headquarters in California launched a GenMEP design add-on to Autodesk which relates to the M&E and plumbing design aspects of BIMs and quickly generates design alternatives, enabling a smooth flow between the building's architecture and services.

Safety and efficiency

The BIM 360 Project IQ was launched by Autodesk. It is a construction management software which uses several data that are connected with machine language (learning) in forecasting and arranging potential risks. It uses data from work sites in the form of audio-visual documentation to provide insights into risk mitigating challenges confronting construction managers. A project IQ pilot developed in partnership between Autodesk and Sinerton Builders was used for monitoring the performance of subcontractors for the enhancement of construction safety. Autodesk has also conducted another version of the software in collaboration with Layto Construction based in Utah and NVIDIA to deploy BIM 360 Project IQ to improve safety on sites. Other AI safety and efficiency equipment such as Compact Assist conducted by Volvo Construction Equipment based in Sweden for recording temperature of an area can be adapted for soil and asphalt mixing and compaction. A SmartTag called VINNIE (very intelligent neural network for insight and evaluation) is a product of Smartvid.com and analyses images and videos through speech and image recognition.

Autonomous construction

Komatsu, a Japanese construction and mining equipment manufacturers in partnership with NVIDIA, has come up with an AI smart construction system to enhance construction site safety. It builds 3D visuals and tracks workers, equipment and other objects on construction sites to ensure safety and efficiency. Over 4,000 project sites in the US have been carried out by smart construction. The smart construction system collects data through cameras which can also be mounted on construction equipment and drones.

There is a proposal by Komatsu to come up with equipment-mounted intelligent cameras which will help such equipment to identify objects on site, thereby preventing collisions.

Monitoring and maintenance

Comfy is an app which can regulate the thermostat of an air cooling or heating system using historical data. The user patterns of requests and preferences are recorded by the application and over time are used to identify future patterns and preferences based on time of day and location. The manufacturers claimed that Comfy uses existing BMS data to regulate internal buildings' temperatures, making it cost and energy efficient.

Doxel, a company in Silicon Valley in the United States, claimed to provide an AI-enhanced software that helps to improve productivity. The equipment consists of robots and drones which are well fitted with cameras and LiDAR (light detection and ranging) sensors for monitoring and scanning work-sites. The device can detect construction errors and take action when it detects deviations from programme schedules, thus saving time and costs.

Muraleedharan (2018) stated that with the emergence of smart buildings, AI will help to ensure an efficient use of energy while roads and buildings will be constructed with intelligent materials. Building materials are now being built from natural earth-friendly materials that are helping to cut costs, assisting in protecting the environment, and helping to conserve resources. Sarath also stated that the use of AI in power transmission infrastructure could help regulate and control wasteful and unplanned power distribution, thereby reducing GHG emissions. He further stated that the advent of autonomous vehicles would help to reduce carbon emissions.

Challenges of artificial intelligence in sustainable construction

According to Anuja Lath (2018), AI has some challenges amongst that are:

- Lack of trust: AI is driven by complex scientific and technical algorithms that are difficult to understand. People who find them difficult to comprehend often abandon its use.
- There is a shortage in the number of skilled professionals who can handle AI technology.
- There is the fear by human staff that AI will displace them.
- It is an expensive technology to invest in.
- It is difficult to lay blame on someone or identify the cause of mistake should software malfunction.
- It has limitations. Not all tasks can be carried out by AI; therefore, it cannot replace all human efforts.
- There are high expectations of AI. Not everyone has a detailed understanding of how it works. There seems to be too much fuss about the technology.
- Prediction failure could have a negative consequence on costs and safety that could discourage the use of AI technology.

Solution to the challenges highlighted

- To create trust in AI, adequate communication and creating awareness about the technology are necessary.
- Humans desire a freedom of choice. Therefore, potential users of AI should be encouraged to explore the technology rather than forcing it on them.
- Construction managers or project managers should instigate active participation of their staff in the use of AI.
- Managements of construction organizations that use AI should attempt to encourage every member of staff in decision-making that engages the use of the technology.
- It is time to change the way we talk about AI in order not to overhype its usefulness that may reduce trust in it.
- The governments can assist in making pecuniary resources available to procure the technology.

Conclusion

We live in a world in which there is an increasing demand for AI systems. There is an increasing need for precision equipment to solve environmental challenges. AI technology is steadily becoming a part of our daily lives. It has advanced the sustainability of the environment in which we live. AI helps to reduce construction costs, saves time, increases productivity, aids the planning, and design of construction projects; ensures safety and efficiency; and above all, assists in advancing the sustainability of the environment. It is therefore necessary for researchers to ensure transparency, fairness, and trust in data obtained through AI systems and platforms. This is necessary to ensure continued use of AI systems for sustainable development. The use of AI is becoming prominent among large construction organisations in the enhancement of workers' safety. There is consensus in construction industries in Canada, the United Kingdom, and the United Arab Emirates on the ability of AI to address safety risks and inefficiencies. It is hoped that manufacturers of AI will work to improve its efficiency in construction processes such as planning and monitoring during construction.

References

Adio-Moses, D. & Asaolu, O. S. (2016). *Artificial intelligence for sustainable development of intelligent buildings.* Paper presented at the 9th CIDB Postgraduate Conference, February 2–4, 2016, Cape Town, South Africa.

Bharadwaj, R. (2019). *AI applications in construction and building – Current use – Cases.* Retrieved from https://emerge.com/AI

Environmental Protection Agency (EPA). (2016). *Green building.* Retrieved from https://archive.epa.gov/greenbuilding/web/html

Jennifer F. M., (2021). Artificial intelligence for good global summit welcomes new frontier for sustainable development. ITU Report 2017, June 7-9, Geneva, Switzerland.

Kurzweil, R. (1990). *The age of intelligent machines.* Cambridge, MA: MIT Press.

Lafargeholcim Foundation. (2019). *Understanding sustainable construction.* Retrieved from https://www.lafargeholcim-foundation.org/about/sustainable-construction

Lath, A. (2018). Six challenges of artificial intelligence. *BBN Times.* Retrieved from https://www.bbntimes.com/en/companies

Luger, G. F. & Stubblefield, W. A. (1993). *AI: Structures and strategies for complex problem solving.* Retrieved from https://www.wikipedia.org

Muraleedharan, S. (2018). *Role of artificial intelligence in environmental sustainability.* Retrieved from https://www.ecomena.org

Newton, J. (2018). *Artificial intelligence in the construction industry.* Retrieved from https://www.curriebrown.com/media/1666/thought-leadership-artificial-intelligence-by-jeremy-newton.pdf

Pitt, M., Tucker, M., Riley, M., & Longden, J., (2009). Towards sustainable construction: promotion and best practices. *Construction Innovation*, 9(2), 201–224. https://doi.org/10.1108/14714170910950830

Rauch, C. (2018). *How artificial intelligence can help sustainable development.* Retrieved from https://medium.com

Russell, S. J. & Norvig, P. (2009). *Artificial intelligence: A modern approach*, 3rd edn. Upper Saddle River, NJ: Prentice Hall.

Rao, S. (2019). *The benefits of AI in construction.* Retrieved from https://constructible.trimble.com/construction-industry/the-benefits-of-ai-in-construction

Schalkoff, R. (1990). *Artificial intelligence. An engineering approach.* New York: McGraw-Hill.

Van Dessel, J. & Putzeys, K. (2007). *Sustainable construction: Let's build the future!* Retrieved from https://www.cstc.be/homepage/index.cfm

Xmaterials. (2018). *How AI is transforming the construction and building materials space.* Retrieved from https://medium.com/@teamxmat?

3 Augmented reality and sustainable construction

Introduction

The construction industry has been revolutionised by computers through practices and methodology, thereby providing the opportunity for efficiency and an avenue for effectiveness (Escamilla & Ostadalimakhmalbaf, 2016). The continuous demand for information, collaboration, and evaluation of processes calls for need of new technologies in construction and engineering (Rankohi & Waugh, 2013). Cleaveland (2011) highlighted that engineering in construction is on the verge of essential innovation in efficiency, standard performance, and safety habits through emerging information technology abilities. Construction firms have begun to adopt information technology in their dealings to increase efficiency and productivity and in order to compete with the diverse increasing complexities of urbanisation (Alshawi, 2007).

Augmented reality (AR) is one of these information technology abilities that conjures virtual objects or models and superimposes them on real environments, providing information and data that are needed in actual time (Hammad, Wang & Mudur, 2009). Wang and Dunston (2007) explained AR as a technology masterpiece that displays virtual information from a computer and inserts this onto a real-time environment that the user can view through sensory wearable devices. AR has benefits for architecture and the engineering field as it consists of live imitation in a real-world environment, interpolating it with real coordinates to give a visual of what is to happen (Behzadi, 2016). It gives the observer a detailed interaction between the real-world and a virtual model, providing the construction details of building by juxtaposing the built as-planned and built-as work of the project (Shin & Dunston, 2010). AR is the type of technology that projects an improved version of a piece, element, or object to a user using real-time information from the immediate environment through sensory input in smart glasses, smartphones, or wearable devices.

Though AR is mostly touted by people as being a futuristic innovation, there are some early examples of its application in 1990s in military fighting. The air force and the navy used it in aircraft and submarines' missiles launching. It displays real-time information concerning the target environment, their altitude, and direction based on precise coordinates, identifying enemies and target

DOI: 10.1201/9781003179849-3

through glass vision display. Over the passage of time, companies and universities began researching ways to develop the technology into a generally useful product by creating AR devices, either as wearable devices or as smart glasses.

This led to MIT's launching SixthSens in 2009. This gadget is useful with smartphones that comprise a projector, mirror, and camera. There was another device created by Google in 2013 in which they incorporated AR devices inside a glass goggle. The smart glasses display information pictures and respond to voice commands. Recently a smart reality mobile AR app was launched by JBKnowledge that gives 3D view of a project according to user specifications. The most recent are the Microsoft HoloLens, the DAQRI Smart Helmet, GAMMA AR, WakingApp, and Arki (Souza, 2019). AR is a new technology gathering the necessary momentum to surpass the usefulness of its cousin, virtual reality, in construction industry. Virtual reality provides users with a digital experience without actual or physical data and information about the surroundings while AR inserts users into the existing plan of the building giving precise measurements, and actual or physical data and information about the surrounding of the building (Izkara et al., 2007). It enables quick removal and necessary adjustment by the end-user or client through wearable glasses that relay superimposed computer-generated images, thereby saving stress, time, and cost attributed to changing design in the construction stage. The wearable devices may come in the form of smart watches, contact lenses, and helmets. It follows some standard requirement for it to be applicable to AR as it must possess a high capacity of information storage, high sensory interface, a communication framework to connect to other devices e.g. Bluetooth, infra-red, and WiFi, and a strong backup plan for its battery system (Shikre & Salamak, 2018). These wearable devices bring graphics and effective sound systems for the information display to the users of AR.

Sustainable construction (SC)

Adetunji et al. (2003) and Opoku and Fortune (2011) agreed that in defining sustainable construction, it has been defined with numerous interpretations as it relates to different sectors. This is also supported by Parkin (2000) who explained sustainable development as the basic background used in processes infused into construction. Sustainable construction propel the construction industry not only towards sustainability but also towards a socially economic environment taking into account the cultural issues along with it (Shafil, Armon & Othman, 2006).

Research studies by Plessis (2002), Baloi (2003), and Abidin (2010), among others, explained the four columns of sustainable construction directed in line with sustainable development as the following:

- Technical
- Economical
- Social
- Environmental

Working principle of augmented reality

The camera at the back of the phone captures information in a traditional manner by placing the camera over the drawing plan. The computer or the phone sees the information captured and the pattern encoded in the drawing and transmits it in real-time information and real-world coordinates, producing a three-dimensional image for the viewer that can be tilted and zoomed through closing in the camera to the plan (Ronald et al., 2001). AR systems need to calibrate cameras in order to mix virtual models and real scene or information together so that virtual objects can behave in a plausible way. AR needs optical camera calibration enforcing physical interaction and constraints between real and virtual objects (Klinker, Sticker & Reiners, 2001).

Calibration of the camera means inserting the precise information such as the height and position of buildings, bridges, waters, or hills into a model, and the model is therefore entered into the camera by entering proper focal height, aspect ratio, centre of positions, orientation, and lens focus. The first approach is to augment compiled video sequences of to-be built that comprises drawings and a 2-D model. This compiled video sequences are viewed with offline interactive calibrations to get a camera positions. These calibrations will enable the video to be performed live (Bajura & Neumann, 1995).

The second approach deals with live streaming of the compiled video sequence calibrating and augmenting the images as mentioned either in optical way or hybrid (Kutulakos & Vallino, 1996). Photo augmentation is where photographs of the design to be constructed are collated. These collated photos are augmented for show to convince the public of the importance and benefit of the design (Koller et al., 1997). Although this process is very tedious, it has a challenge of not being able to alter the design when showing it. The augmentations can be altered at different speeds.

How augmented reality works

AR operations and functions have been the main focus during its introduction. It is perceived to be of certain range data that encompasses various components functioning together to form an idea of reality and expressing the idea in computer vision. It fits into various reflective surfaces such as the following:

- Head-mounted displays
- Mobile phones
- Glasses
- Handheld devices
- Screens

Its technology is expressed by involving in-depth tracking that uses a sensor data in calculating the distance of an object, simultaneous localization and mapping (SLAM), and other components such as the following:

Cameras and sensors

The data collected from the user's interactions are sent for processing. The cameras scan the environment and the surroundings with directed information, and such information inputted locates a physical object and produces a 3D model of the object captured. Common smartphone cameras for taking pictures and recording videos could be used or a specially designed duty cameras such as the Microsoft HoloLens.

Processing

There is little difference between the operations of the AR to those of the smartphones used presently. It acts as a little computer system of its own even though it works better and more efficiently by using a central processing unit (CPU), flash memory, random access memory (RAM), and other compatible connections such as Bluetooth, Wi-Fi, or GPS in assisting and facilitating a better angle, orientation, speed, direction, and movement.

Projection

The projection of AR is quite simple as it integrates something as simple as a sensor and relates that to the data collected or to be collected to project digital images or content for proper usage. This projection has not been the most popular as most firms or companies have not yet fully integrated it into their system of operations.

Reflection

AR uses objects like that of a mirror to widen the scope of the human eye from different perspectives. It is arranged with arrays that come in various shapes and sizes; some with a small curved surface (mirror) while others have a double mirror size to reflect illumination into the camera directed to the user's eyes with the aim to create a reflection path which in turn propagates accurate images that are aligned according to the system set.

Augmented reality in today's world

What the world sees and experiences are the outcome of using technology in enhancing and creating more information from texts, images, and sound for a more prolific experience than ever than before.

AR is the output of using technology to superimpose information such as sounds, images, and text for the world we see and experience. Picture the 'Minority Report' or Iron Man' style of functionality and interactivity, it differs from the setup of virtual reality. Virtual reality explains the environment produced by a computer for better interaction and knowledge. AR creates a better picture of the reality seen by adding extra information rather than obliterating it.

AR is not as it was in the early development contrary to the opinion that it is just being propounded as a futuristic technology. The heads-up displays present in fighters' planes have been present dating back to the 1990s. This aids in displaying information and attitude, speed and direction, and years later, due to upgrading, was able to highlight objects in the target field. Over the decades, many laboratories and companies have made devices that contain AR. Out of which in 2009, MIT Media Lab's Fluid Interfaces Group presented SixthSense, a device that brings into use the functionality of small projectors, cameras, and that of smartphones and mirror. The device is strapped onto the neck of the user in a lanyard style. The sensor devices (four of them) are manipulated to the image that is emanating from the SixthSense.

It is necessary for technology to be reduced in size but with more functionality. This aspect drove Google to making virtual reality, something that is portable for more efficiency in December 2015. They made it in the form of spectacles as they are easier to wear and convey around. Several features were attached as the glass acts like a small projector that is subjected to whatever command the user gives in accordance with the programmed function.

Gaming has gone global. People tend to play games to relieve themselves of the stress that has been accumulated during the day's work. The game Pokemon Go created in 2016 with the use of AR became something that amazed the world with an almost total acceptance. According to the survey carried out by CNET, the users estimated to be involved in the game numbered more than 100 million at its peak. The special feature of the game is the interaction as it allows its users to see characters programmed in the game on their own. The objective was to create an avenue where some would fight others and some would pocket monsters within a locality in the AR designated gym. In a bid to make enhancement to the components of AR, the researchers are planning on adding holograms to be used in VR as this will put it more out there for people to participate at once since the audience that hologram can relate to are many.

Augmented reality devices

AR is significantly supported by contemporary devices. Some of these are as follow:

- Smartphones
- Tablets
- Handheld devices or Google glasses

The most practical explanation is that these technologies continue to evolve because of their several uses in the industry. Even though most of the components are mostly hardware, they also comprise other components that cannot be touched such as digital compasses, GPS, accelerometers, displays, sensors, gyroscopes, cameras, CPU, and other parts that make up the AR.

Devices suitable for augmented reality fall into the following categories

Mobile devices (smartphones and tablets)

There are numerous ways by means of which AR can be used to influence our lives. One such way is the use of smartphones and sophisticated tablets. Their function ranges from sports, entertainment, and gaming to analysis of different sorts and even all forms of social networking. These devices are compatible with the usage of AR as it is very convenient to convey transparent data displayed from the head-up displays (HUD) without restriction and channel that directly to the user's perspective. With its acceptance generally increasing, it has grown from the military use by their fighter pilots to more use in industries, sport, aviation, management, and manufacturing.

AR glasses (or smart glasses)

In putting units together that make up the parts of AR, notifications and other displayable functions could be displayed through Laforge AR wear, Google glasses, Laster See-through, and Meta 2 glasses. They are also directed at aiding assembly line workers to have access to hands-free contents. All this could be made possible from smartphones of any size and shape. This in conjunction with smart lenses is making AR very accessible.

Samsung and other manufacturers such as Sony have implemented AR lenses in their development. Though they are both manufacturing companies, Samsung is directing its resources towards smartphones that have lenses that the phone could infuse with its functions. Sony, on the other hand, is working towards separating lenses from AR devices by also incorporating extra features in the storage of data and for taking photos from the same device.

Other potential areas for AR include the following:

- Education
  It could be used interactively in subjects such as Chemistry, Mathematics, and other science-related subjects for training personnel who would be able to teach by means of models that can easily be interacted with.
- Medicine and healthcare
  The application of AR could help health practitioners to diagnose more effectively, train workers, and monitor the patients, the staff, and other related medical fields.
- Military
  Military personnel could use AR for detecting, advancing and close objects, marking some objectives to suit a plan, and also applying it in advanced navigating positions.

- Tourism
  In terms of a tourist centre, several attractions could be viewed and enjoyed with the aid of an AR implemented system. It could help in giving directions to tourists, in terms of sightseeing or viewing objects or artefacts of significance. It also leads to smoother navigation and the collection of data collected on arrival at their destination.
- Broadcasting
  For superior delivery of the broadcast programmes, AR could enhance the streaming of events and other related activities. It does this by overlaying contents selected.
- Industrial design
  In order to obtain a more vivid version of a design, the application of AR in designing, visualizing, constructing, and calculating these models to aid better and swifter industrial design.

Types of augmented reality

Marker-based

The special features of this type of AR can be accessed through the use of camera to scan appropriate areas of the work. It can function as image recognition as it can be viewed by specified virtual objects and cameras. A quick response (QR) code could be used in its application as well as special signs associated with it. Most of the process includes the calculation of the orientation and an appropriate marker to direct a possible location. This marker changes the images obtained to a model in the form of a 3D representation by integrating animations for the user to view and access it.

Markerless AR

This is also referred to as location-based AR. It utilizes a gyroscope, GPS, accelerometer and compass to showcase the operator's or user's whereabouts. The data obtained will then determine what AR content to navigate and what to do with such data related to a specific location in time. The availability of smartphones inputted in this type of AR will produce maps and directions coupled with other information about nearby businesses.

Projection-based AR

Projection-based AR emits artificial light to physical or visible surfaces and permits interaction with it and the corresponding components. This application is mostly used in science fiction films at home or in the cinemas. A well-known example was Star Wars. The AR detects the way the user is going to relate with a projection by acting on its alteration.

Superimposition-based AR

This type of AR replaces the augmented view of reality to that of the original, whether fully or partially. The system must be able to recognise image because without the proper capturing of the image, the concept is of no value. The IKEA catalogue application permits users to use items of furniture in rooms, showcasing a typical superimposed AR.

Smart City planning and building: Augmenting the possibilities

The consistent use of AR in the construction industry will propel its development towards planning and executing smart cities and buildings for swift, sustainable construction with the best cost control available. The huge amount of data collected by cities could be made visible by the use of AR. Using AR, professionals such as architects could find more expression to showcase creative models in a straightforward way. This gives them a better view of what the urban planners could incorporate in their allocation of spaces and designs as well. This is realised by engaging visualisation to widen the nature of the work at hand, thereby helping them in making a first-hand decision. Whether the designers are trying to construct a skyscraper, no matter the location, or whether they are using it to monitor or construct transportation centres, using AR makes projects easier, more profitable and more manageable.

Augmented reality in construction

There are various areas of application of AR in construction and they are summarized in Table 3.1 and Figure 3.1. Table 3.1 indicates the areas and their respective codes while Figure 3.1 represents the relationship of the areas in the consideration.

Scheduling and project progress tracking

In the schedule of work to be done before and after the construction stages, the progress of the work should be in line with the programme of work that the

Table 3.1 Augmented reality in construction

Code	Meaning
ARA1	Scheduling and project progress tracking
ARA2	Communication and data acquisition
ARA3	Quality and defect management
ARA4	Time and cost management
ARA5	Employee training and safety management programme
ARA6	Man-labour hours

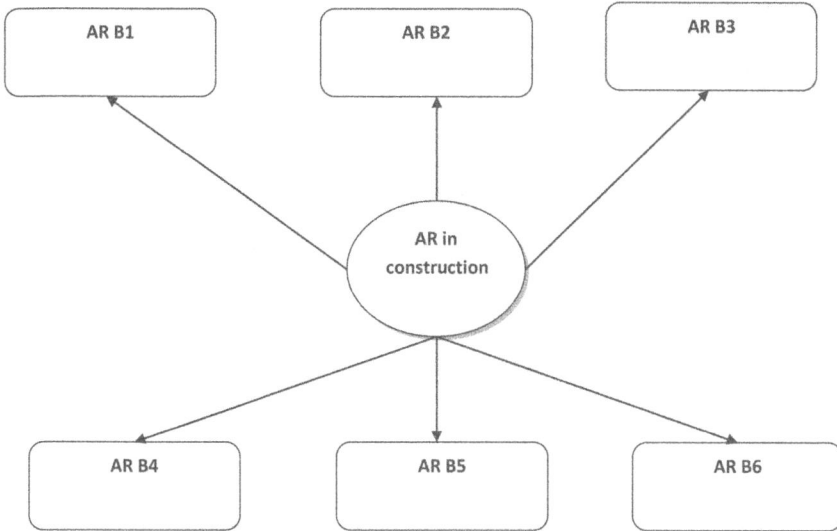

Figure 3.1 Augmented reality in construction.

designers decided on from the beginning of planning the project. AR can esti‑ mate the various works to be done in respect of the proposed work planned. It also gives an annotated visualisation of how the work is progressing. By using AR on tablets, PCs smartphones or other mobile configured devices, it has agreed to be far more effective than Gantt charts or other 3D‑related models (Meza et al., 2015). According to Wang et al. (2013a) and Kim et al. (2013), the consist‑ ent application of AR in construction paves a way for comparisons between the progress of the project and that of the newly established work.

Communication and data acquisition

A successful project starts from the small details in communication between the professionals and the artisans involved in the work. The obtained data are processed to use in ensuring the progress of the project. Pejoska et al. (2016) stated that obtaining project information while on site, along with effective dissemination of information among the professionals, is enhanced with the introduction and implementation of different AR systems as compared to the crude way of collecting the available information. AR technologies can be used for obtaining primary or secondary field data in construction. It also denotes the modes of communication in terms of responsibility between pro‑ ject participants and project information flow for a better project. In taking corrective measures in minimising delays that might have been caused by cost and other factors, AR points to the project manager to take action that will at the end of the day reduce any incompetence in performance (Bae, Golparvar-Fard & White, 2013).

Quality and defect management

Quality management and defect control are as important as any part of construction management. The client's needs are always considered in order to deliver a project of standard quality for the satisfaction of the client or the professionals in charge. By introducing automation and the implementation of defect management into construction, AR plays an important role in global construction by showing the process of marker-based AR technology discussed earlier for defects and quality management.

The process of defect management by AR that was reviewed by Kwon, Park and Lim (2014) referred to the AR functionality on site-experiment by using mobile devices of the type chosen.

Time and cost management

Time and cost are undoubtedly the major constraints in any construction process. The application of numerous management sectors (risk, value, and material) in construction projects is to maximise construction project operations in order to save expenses. Rather than wasting unnecessary time and energy in going through a lengthy process, AR has laid a systemised platform for the professionals in construction to minimise the time and the cost while still obtaining a better quality at the end of completion. This was affirmed by Wong, Wang, Li and Chan (2014) who added that the features help the project managers or any professional in charge of the project to reduce work. Most of the activities are carried out by the AR technology implemented at the onset of the project which in turn gives a nearly (if not perfect) job. Instead of the unnecessary or unaccounted wastage of labour and materials, the AR gives a platform that provides better service to management within the shortest possible time.

Employee training and safety management programme

A safety management system is concerned with making protocols available that could help to safeguard the health of the workers, while also the materials on or off site. Several workers have died over the years. These fatalities were caused either by failing to train employees on matters of safety or by neglecting the so-called small safety protocols. Training of employees is an important aspect of every construction project. The activities are not easy to perform at a desired or standard level. However, AR technologies can provide the employees with the necessary adequate training by implementing safety management system as a required specification. AR can be used to train workers pertaining to construction with either participation in light or heavy construction equipment (Chi, Kang & Wang, 2013). Wang et al. (2007) supported this by suggesting that this technology can be used to train workers to operate construction equipment required within the scope of construction.

Man-labour hours

When professionals really understand the benefits that are attached to the usage of AR and other related technologies in construction, the more the industry will progress. In terms of man-labour hours, AR can allocate the unit work per unit worker. This saves the consultant or the project manager in charge a great deal of time and effort. The efficiency of the available man power is also utilised in the best possible way. Also, in synchronising a particular role with the time of work, the technology provides consistent and accurate measurement of the rate of work done. From there resources can be allocated to where they are needed. This tends to save money and reduce wastage of resources.

Benefits of augmented reality for construction project

The accuracy of construction projects

In order to achieve a high degree of accuracy, the constituents should be well positioned to align with an identified system or goal. The construction industry seeks more accurate results at the completion of a project. This has driven it to accept technologies that could aid in removing errors and defects. Whether they are errors seen detected early on or encountered later, the application of AR will provide an accurate platform to address these because every action taken is calculated and weighed.

Saves money and time

It is well known that time and costs are the major reasons that professionals use the application of AR and its systems. The utilisation of the various systems embedded in the set-up ensures that works are properly saved, timed, and allocated while also providing a greater way of minimizing cost when the functions are considered that it is going to perform throughout the construction stages. It allows for alterations of several components of the building such as the walls and easily replaces them with an alternative. The engineer or operator in charge needs to be conversant with the functions of the technology and how to use it to make corrections to the identified problems. Moreover, it assists the architect by showing a better 3D model of the interior of the building with the information provided. It facilitates the work of the architect who is able to see the design for what it is in actual time.

Construction project management

Apart from the advantages in terms of time and cost effectiveness, another benefit of AR is aiding in project management of buildings that are being constructed or those that have been completed. Actual time realisation removes every bit of doubt (probability) in the work as the professionals involved can easily access modern and accurate information rather than the manually recorded estimates and measurements that were available to them prior to the introduction of AR. Designers of any discipline can also showcase their creativity and realistic ideas through the segmented components available in the use of AR.

Design analysis for designers and architects

This is another way in which AR assists designers in producing better and most realistic ideas in construction. The designs created through whatever platforms are suitable are easily analysed through the components of the technologies available. This gives the designers an opportunity to make corrections because they now have an in-depth analysis of the designs before commencing the construction. In cases where an omission was discovered, the designers can go back to the design and alter it immediately. The technology will provide a smooth surface for interchangeability that is always an advantage when working on technology-enhanced plans.

Actual construction of a building

Even before the start of the project, AR offers a swifter rate of the actual construction of a building by theoretically analysing every section of the work that is to be done. This enables the professionals to have more than a basic idea of the project at hand and how to go about it. Therefore, the building can be viewed in its totality thanks to the models that the AR provides.

Scheduling

AR is important in allocating and relating a programme of works. It enables the client, designer, and constructor to visualize and interact with the as-planned drawing, effecting the changes needed before the actual construction takes place (Zollmann et al., 2014). Meza et al. (2015) in their research on ranking the benefits of work scheduling or programming among AR, 3D or a Gantt chart, found that the respondents indicated that AR is highly effective and more beneficial than the two other options. Wang et al. (2013b) were also of the view (2015) that there is effective and efficient progress monitoring on work schedules through AR. It also ensures faster update of work plans after some design changes (Klinker, Sticker & Reiners, 2001). In addition, AR helps construction teams to be versatile by familiarising themselves with work details in scheduling before construction in order to save time and avoid rework (Chandarana, Shirke & Desal, 2013). Application of AR technology encourages efficient planning, scheduling, controlling, and monitoring in construction projects (Khalid et al., 2013).

Information retrieval

There is no doubt that information mechanism and retrieval are vital elements in the success of a construction project. AR plays an important role by giving easy access and ensuring a faster mode than the traditional approach to project participants and the relevant stakeholders (Pejoska et al., 2016). Bae et al. (2013) emphasised the need for first-hand information that aids professionals and clients to identify and correct avoidable errors, thereby reducing cost from rework and delays. AR presents an avenue to display project information such as drawings and measurements that are needed on site according to the user's location by visualising it (Borkin et al., 2016). AR helps in checking of building details in the construction stage through wearable devices or the use of smart glasses are connected to computers that contain virtual animations, drawings, and contract specifications. Gelder et al. (2013) explained AR as an integrator of digital information with user's information in real-world time, allowing human senses to interact with the experiences.

A navigation feature will be added to the smart glasses and the ability to access information on the Internet if necessary. This will also help the designated site managers to check the quality of work to be done with what was on-going on site and produce real-time data of the work if necessary. Smart glasses usage in connection with AR will help staffs to exchange communication on site through emails and short messages that are accessible on the smart glasses or wearable devices platform because of their connection to the Internet. In other words, AR helps to establish innovations for construction workers on the conditions of the site according to their duties and authorisation. AR saves time and is cost-effective in the sense that reworks are avoided through connection to the Internet of Things. This displays the drawings and contract specifications in a real-time environment for staff instead of their having to go back to the office to check it out.

Design and marketing

Automated reality integrates virtual pictures into images of real life. It shows clients the real-time images of what they want before construction. This further convinces the client in terms of commitment to the project (Klinker, Sticker & Reiners, 2001). The capability of AR to bring virtual models into real-life interaction can guide a designer or client to identify problems that might be encountered in the construction stage or in the functionality of the design, thereby avoiding delays and rework at the construction stage (Dünser et al., 2007).

AR provides an option of seeing through the whole building from the design stage before construction in order for the designer or client to experience the aesthetics, beauty, and workings of the project (Chandarana, Shirke & Desal, 2013). This is done through superimposing a 3D model on the drawings of the project. It provides an avenue for designers and developers to integrate their design into real-world data and to evaluate its function before presenting it to clients and investors for their satisfaction and approval. They are able to visualize the design and decide whether it is according to their specifications. The developers are thus able to effect any changes the client desires before construction commences (Khalid et al., 2013).

Renovation and maintenance

AR shows information about some work embedded inside a wall i.e. a conduit system such as where pipe works follow in the wall or throughout the building so that the engineer can informed at the speed required to complete the maintenance or renovation within the specified time (Azarbayejani & Pentland, 1995). Experts in the construction industry find it easy to use AR in carrying out difficult maintenance and repair works in construction projects through interfacing the real-time graphics with the real environment (Chandarana, Shirke & Desal, 2013).

In summary, the application of the principle of AR in construction will overall sustainability of the construction industry. These benefits are expressed in Figure 3.2.

Challenges of augmented reality

For a large construction project where the drawings are large and the information is vast, a special processor will be necessary that can translate the heavy information more quickly into two- and three-dimensional models in the actual sense. To do this requires a fast computer infrastructure, a 5G mobile network and large amounts of storage in terabytes. A considerable backup plan will also be necessary for AR to be implemented, as well as software for sieving the useful information from that which is not required to aid better performance of AR technology (Chandarana, Shirke & Desal, 2013).

Its ability to interface with real-time information gathering is another challenge where some devices attached to AR such as camera cannot function properly in an outdoor environment.

AR technology requires precise and detailed geographic information about the real environment in order to place the current camera position to augment the synthetic information known as a virtual model. Natural objects and resources are depicted such as rocks, hills, water, and trees. It is challenging to make a model of a site containing one of these natural resources. Buildings are renovated, demolished, and reconstructed every day while fluctuating weather and seasons may inhibit access to gathering actual information about the site or the construction environment of the built area. There could also be tracking and alignment

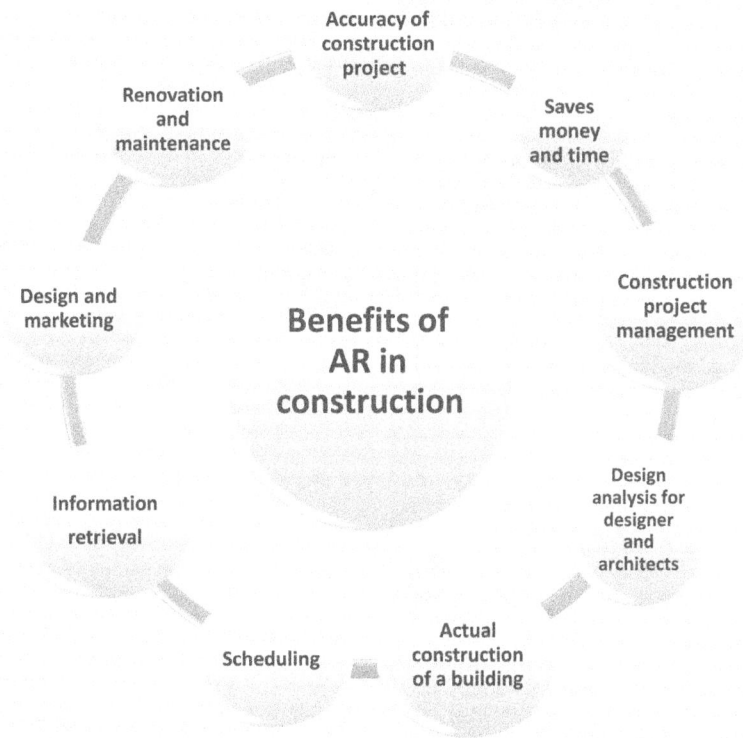

Figure 3.2 Benefits of AR in constriction.

difficulties. There is less usage of AR on site; it is mostly used in offices to view and simulate the design of the project due to its tracking technological complexities and camera orientation alignment. Nonetheless, it is used in the areas of project progress tracking and defect detection (Rankohi & Waugh, 2013). AR requires GPS for tracking that can only be effective within a 30-cm interval (or radius) and they cannot work perfectly in indoor operations. They tend to suffer from occasional fluctuations in the GPS signal (Perey & Graziano, 2013).

Conclusion

The ever-increasing need of the client has brought some changes to the construction industry in terms of the expected output that is expected to flow seamlessly and swiftly. Herein lies the source of the satisfaction to be derived, whether short or long term. AR brings into play facets that are not easily identified as AR encompasses various segments to make construction easier. Incorporating sustainability through the adoption of AR will increase the durability of the construction and the way the environment is viewed as a more vivid depiction of the section of work to be carried out is available, from the onset to the completion of the project.

References

Abidin, Z. (2010). Investigating the awareness and application of sustainable construction concept by developers. *Habitat International*, 34, 421–426. http://dx.doi.org/10.1016/j.habitatint.2009.11.011

Adetunji, I., Price, A., Fleming, P. & Kemp, P. (2003). Sustainability and the UK construction industry: A review. *Proceedings of the Institute of Civil Engineers, Engineering Sustainability*, 156(ES4), 185–199.

Alshawi, M. (2007). *Rethinking IT in construction and engineering: Organizational readiness.* Abingdon: Taylor & Francis.

Azarbayejani, A & Pentland, A. (1995). Recursive estimation of motion, structure, and focal length. *IEEE Pattern Analysis and Machine Intelligence*, 17(6), 341. http://www.white.media.mit.edu/vismod

Bae, H., Golparvar-Fard, M. & White, J. (2013). High-precision vision-based mobile augmented reality system for context-aware architectural, engineering, construction and facility management (AEC/FM) applications. *Visualization in Engineering*, 1(1), 3.

Bajura, M. & Neumann, U. (1995). Dynamic registration correction in video-based augmented reality systems. *IEEE Computer Graphics and Applications*, 15(5), 52–60.

Behzadi, A. (2016). Using augmented and virtual reality technology in the construction industry. *American Journal of Engineering Research*, 5(12), 350–353.

Borkin, M. A., Bylinskii, Z., Kim, N. W., Bainbridge, C. M., Yeh, C. S., Borkin, D., Pfister, H. & Olivia, A. (2016). Beyond memorability: Visualization recognition and recall. *IEEE Transport Visual Computation Graph.* 22(1), 519–528.

Chandarana, S., Shirke, O. & Desal, T. (2013). Review of augmented reality applications: opportunity areas and obstacles in construction industry. *Retrieved from:* http://review-of-augmented-reality-applications-opportunity-areas-amp-obstacles-in-construction-industry

Chi, H.-L., Kang, S.-C. & Wang, X. (2013). Research trends and opportunities of augmented reality applications in architecture, engineering, and construction. *Automation in Construction*, 33, 116–122.

Cleveland, A. B. Jr. (2011). Emerging tools to enable construction engineering. *Journal of Construction Engineering and Management*, 137(10), 836–842.

Dünser, A., Grasset, R., Seichter, H. & Billinghurst, M. (2007). Mixed Reality User Interfaces: Specification, Authoring, Adaptation, at 2nd International Workshop IEEE Virtual Reality 2007 Conference, Charlotte, North Carolina, USA.

Escamilla, E. & Ostadalimakhmalbaf, M. (2016). Capacity building for sustainable workforce in the construction industry. *The Professional Constructor*, 41(1), 51–71.

Gelder, D., Zak, C., Lord, T., Greiner, M., Notova, N. & Barrena, M. (2013). Inside AR. *Augmented Reality Magazine*, 3, 1–16.

Hammad, A., Wang, H., & Mudur, S. P., (2009). Distributed augmented reality for visualizing collaborative construction tasks. *Journal of Computing in Civil Engineering*, 23(6). https://doi.org/10.1061/(ASCE)0887-3801(2009)23:6(418)

Izkara, J. L. Perez, J., Basogain, X. & Borro, D. (2007). Mobile augmented reality, an advanced tool for the construction section. *Paper presented at the proceedings of the 24th W78 conference*, Maribor, Slovenia.

Khalid, C. M. L., Mohamed, Z., Fathi, M. S., Zakiyudin, M. Z., Rawai, N. M. & Abedi, M. (2013). *The potential of augmented reality technology for pre-construction.* Proceedings of the 2nd/2013 International Conference on Civil, Architectural and Hydraulic Engineering (ICCAHE 2013), Zuhai, China, 27–28 July, 102.

Kim, C., Park, T., Lim, H. & Kim, H. (2013). On-site construction management using mobile computing technology. *Automation in Construction*, 35, 415–423.

Klinker, G., Sticker, D. & Reiners, D. (2001). *Augmented reality for exterior construction applications in augmented reality and wearable computers.* In: W. Barfield and T. Caudell (Eds.). Lawrence Erlbaum Press.

Koller, D., Klinker, G., Rose, E., Breen, D., Whitaker, R. & Tuceryan, M. (1997). Real time vision based camera tracking for augmented reality applications. *Proceedings of the ACM symposium on virtual reality software and technology (VRST)*, Lausanne, Switzerland, 87–94.

Kutulakos, K. N. & Vallino, J. (1996). A fine object representations for calibration-free augmented reality. *IEEE Virtual Reality Annual Symposium (VRAIS)*. Santa Clara, CA. https://doi:10.1109/VRAIS.1996.490507

Kwon, O.-S., Park, C.-S. & Lim, C.-R. (2014). A defect management system for reinforced concrete work utilizing BIM, image-matching and augmented reality. *Automation in Construction*, 46, 74–81.

Le, Q. T., Pedro, A., Lim, C., Park, H., Park, C. & Kim, H. (2015). A framework for using mobile based virtual reality and augmented reality for experiential construction safety education. *International Journal of Engineering Education*, 31(3), 713–725.

Opoku, A. & Fortune, C. (2011). The implementation of sustainable practices through leadership in construction organizations. In: Egbu, C. and Lou, E. C. W. (Eds.), *Proceedings of the 27th Annual ARCOM Conference, September 5–7, 2011.* Bristol, UK: Association of Researchers in Construction Management, 1145–1154.

Parkin, S. (2000). Sustainable development: the concept and the practical challenge. *Proceedings of the Institution of Civil Engineers, Civil Engineering*, 138(3–8), 3–8.

Pejoska, J., Bauters, M., Purma, J. & Leinonen, T. (2016). Social augmented reality: Enhancing context-dependent communication and informal learning at work. *British Journal of Educational Technology*, 47(3), 474–483.

Perey, C. & Graziano, T. (2013). *Reality-assisted 3D visualisation for urban professional users.* Retrieved from https://www.slideshare.net/armediaaugmented/augmented-reality-assisted-3d-visualization-for-urban-professional-users

Rankohi, S. & Waugh, L. (2013). Review and analysis of augmented reality literature for construction industry. *Visualisation in Engineering*, 1(9). http://dx.doi.org/10.1186/2213-7459-1-1

Shafil, F., Armon, Z. & Othman, M. Z. (2006). Sustainable building and construction in Southeast Asia. *Proceedings of the Conference on Sustainable Building South Asia, Proceedings of the Sixth Asia-Pacific Structural Engineering and Construction Conference (APSEC)*. Kuala Lumpur, Malaysia.

Shikre, O. & Salamak, M. (2018). Building information modelling and augmented reality in improvement of roadway bridges. *Silesian University of Technology, ul. Akademicka*, 5, 44–100.

Shin, D. H. & Dunston, P. S. (2010). Technology development needs for advancing augmented reality-based inspection. *Journal of Automation in Construction*, 19, 69–182.

Wang, X. & Dunston, P. S. (2007). Design, strategies, and issues towards an augmented reality-based construction training platform. *International Journal of Construction*, 12, 363–380. https://www.itcon.org/2007/25

Wang, X. (2009). Augmented reality in architecture and design: Potentials and challenges for application. *International Journal of Architectural Computing, 7*(2), 309–326.

Wang, X., Truijens, M., Ding, L. & Lavender, M. (2013a). Integrating building information modelling and augmented reality for construction projects in oil and gas industry *International Journal of Architectural Computing, 7*(2), 309–326.

Wang, X., Love, P. E., Kim, M. J., Park, C. S., Sing, C. P. & Hou, L. (2013b). A conceptual framework for integrating building information modelling with augmented reality. *Automation in Construction, 34,* 37–44.

Zollmann, S., Hoppe, C., Kluckner, S., Poglitsch, C., Bischof, H. & Reitmayr, G. (2014). Augmented reality for construction site monitoring and documentation. *Proceedings of the IEEE, 102*(2), 137–154.

4 Big data and sustainable construction

Introduction

Every day, millions of digital data that the administration or companies store for future use are generated globally. This data registers the whole of human activity and also records the natural phenomena that occur in all parts of the planet. Big data is therefore defined as the accumulating, processing, studying, and using data on a large scale. Data collected collaboratively is undoubtedly an important source of information for both the private and public sectors. The construction sector is also one of these sectors that have benefited most from its use. Big data is chaperoning a cogent role in sustainable development from the public and institutional sphere, as it brings along implemented transformation of big data into sustainable data.

Recently, data storage has increased exponentially. The various kinds of data available differ from each other in complexity because not all data are structured to gather information the same way. The value of data capturing, processing, and analysis varies from one sector of application to another. A good example of value added to big data is the case of retail giants such as Walmart or Amazon. They make use of data mining to identify, predict, and react in real time to customers' expectations. In referring to how big data is defined in other industries other than that of construction, readings from previous big data stored could still be used today to make decisions across practices. However, it seems the construction industry in particular is still lagging behind in realising that the implementation and its application in construction activities are vital as it leads to significant savings. However, there might be a difference in terms of the quantity of data to be stored, especially when compared to other sectors of the economy that is heterogonous as most make use of the same data over and over again. As the amount of data grows with the passing of time, the urgency increases to streamline it down to a size that is workable without disposing of any information from it.

In comparison to other industry sectors, one of the challenges that the construction industry has been facing is the low level of productivity. Many factors have been identified by numerous research studies and reports. A project management institute study from 2013 identified the risks for project failure as being due to poor communication. Hence, over 7% of projects cost are always at the risk

DOI: 10.1201/9781003179849-4

of gathering enough information. Therefore, companies are required to take the role of data very seriously, even at the early stage of the construction. Whether the cause is related to a technical issue or human factors, it is ultimately a management issue on how to ensure the project is successfully completed on time, within budget and with the expected quality. Therefore, big data analytics can offer opportunities to address or improve each of the sections identified as a priority. The different inputs in big data allow better levels of certainty about status reports and predictions.

The big data

Since the introduction of much software used for various purposes across different industries, there is the need to explain and to have an understanding of what data is. The definition by Manyika et al. (2011), Tien (2013), and Waller and Fawcett (2013) expressed big data in terms of its size, and they explain its difficulty in accessibility by some software in managing operations related, as it captures a specific target when it is put to use within a duration allocated for it.

The more the need for big data comes into play, the more researchers are coming up with various definitions of their own. Courtney (2012) and Russom (2011) explained the concept of big data by expressing the factors that drive it. These entail variety, volume, and velocity. A further study by Abbasi, Sarker and Chiang (2016) added to the definition by Courtney and Russom by including another factor to the 3Vs (variety, volume, and velocity), namely veracity to expand the previous definition. It can be said that big data is a volcano of data that has its source from unstructured differences with different levels of usage and application. The explanation here is not only how big or much the data comprises, but also that the more the data, the more the difference between them. The basic idea of big data is to generate insights that are unattainable with smaller quantity of data.

Big data in construction

The construction industry is also moving towards the phase of digital resolution where the majority of the activities involved are carried out using technology. The industry is finding a constant increase in data from other industries or disciplines as most of them have already gone digital. The implementation of the CAD information component of the building information modelling (BIM) that aids in capturing a multi-dimensional card system has really helped in bringing the professionals from different constructional disciplines together to work more efficiently.

The major aim here is to have a system or a platform that works towards a better sustainable construction environment both for the client and the professionals alike. The targeted information is expected to provide efficiency and stability for better smart approaches to the information presented over the course of the project. There is also an integration of systems that are expanded in integrated building control and monitoring system (IBCMS), computer-aided facility management (CAFM), and BIM. The integration of these strategies could help in making reasonable, and better making decisions in respect to having a construction that is manageable and sustainable.

The nature of the construction industry is such that projects are temporary and are made up of a team of people who work together until the completion of the project. This has been noted by Lo, Fung and Tung (2006) to be the reason why the construction industry seems to be behind in productivity when compared with the manufacturing industry. They opined that the short-term team work results in lower levels of coordination, planning, and communication when compared to other industries. Some modern equipment used in construction has been designed to generate data. The MIT Technology Review (2015) revealed that drones could be used to monitor the progress of construction. Drones were sent out to patrol the construction site once a day and footage of videos were collected and converted to 3D pictures. These pictures were compared with the original architectural drawings and work methodology after the pictures had been fed into software. This showed the construction managers where there was progress on the project and where delays could occur.

The report of MIT Technology Review in 2015 expatiated on the application of drones to monitor how the construction is progressing by automatically patrolling the work site once per day to collect video footage and data. The recorded image of the site is then converted into a three-dimensional picture of the selected site, which is then transferred into software that compares and computerizes architectural plans as well as the construction work plan, displaying when each element should be completed. The software also shows this to the managers or the operators in charge the progress of the project, and they can then systematically make suggestions where works need to be done according to the projected time line.

Big data engineering and big data analysis

In order to comprehend the differences between all the parts that make up big data, it is necessary to fully understand the two major parts, namely big data analytics (BDA) and the big data engineering (BDE). The major function of the BDE is to provide necessary support in processing the relevant information that is required for activities omitted. BDE comprises technology stacks such as Berkeley Data Analytics Stack (BDAS) and Hadoop in ensuring the function it is expected to perform. BDA is another integral aspect of big data that relates to the tasks responsible for extracting the knowledge in driving decision-making. BDA is mostly associated with the techniques, processes, and principles contained in the big data that is allocated towards a specific system of operation. It is also expected to show where the big data is needed by exposing the pattern encored in the system and pick up needed information from there. Decision-making is made easier through a data-oriented platform that makes operations move faster and better circulated to the participants concerned.

Key principles of sustainable construction

By means of the efficient use of resources and ecologically sound processes, the conservation and preservation of the ecosystems along with maintaining a balance between development and the planet's carrying capacity, a safe environment should be everyone's goal.

The key principle upon which the above statements have been phrased can be summarized in the main principles of sustainable construction. Some models detailing principles of sustainable construction will now be briefly considered.

Hill and Bowen categorized the principles of sustainable construction into four pillars, namely social, economic, biophysical, and technical. These are integrated with a set of overarching, process-oriented principles. These process-oriented principles suggest approaches to be followed in deciding the emphasis to be given to each of the four pillars of sustainability, and each associated principle, in a particular situation as shown in Figure 4.1. Davenport, Barth and Bean (2012) listed the principles of underlying concepts towards a safe and sustainable environment as the following:

1 Environment
2 Equity
3 Public participation
4 Futurity

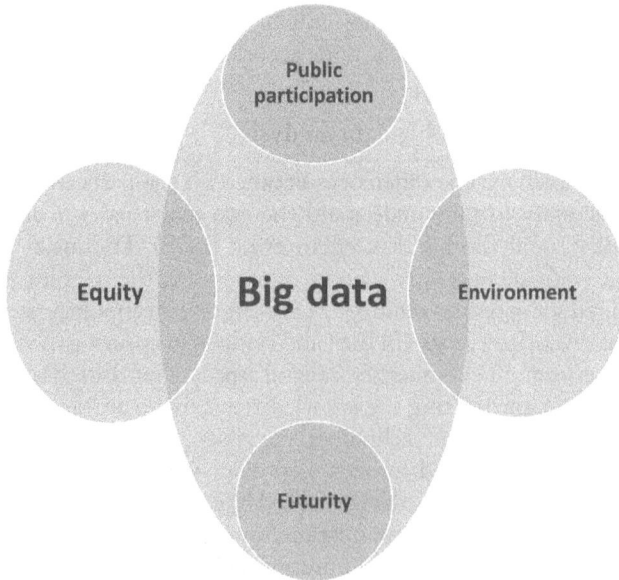

Figure 4.1 Pillars of sustainability in big data.

Conclusion

Big data would improve the sustainability of construction projects, leading to improved projects that are environmentally responsible and take the safe environment of future generations into consideration. However, the use of big data for sustainable construction would mean that a cost would be tied to the calculation of each data point and for construction organizations with massive data centres, this may lead to increased costs. With the drift towards open sourcing data centres and the use of cloud services, it means that more people would benefit from the data from sustainable construction. This would lead to a more sustainable environment as data from different industries could be combined to make more accurate predictions. The use of big data in green buildings would lead to the adoption of wireless networking technologies and better secured devices. It would also increase IT connectivity of devices. Design standards would be more empirical and will impact sustainability. This would provide economic value.

References

Abbasi, A., Sarker, S. & Chiang, R. H. L. (2016). Big data research in information systems: Towards an inclusive research agenda. *Journal of the Association of Information Systems*, 17(2), 1–32. http://aisle.aisnet.org/jais/vol17/iss2/3

Courtney, M. (2012). Puzzling out big data. *Engineering and Technology*, 7, 56–60.

Davenport, T. H., Barth, P. & Bean R. (2012). How 'big data' is different. *MIT Sloan Management Review*, 54(1), 1–5.

Lo, T., Fung, I. & Tung, K. (2006). Construction delays in Hong Kong civil engineering projects. *Journal of Construction Engineering Management*, 132(6), 636–649.

Manyika, J., Chui, M., Brown, B., Bughin, J., Dobbs, R., Roxburgh, C. & Byers, A. (2011). *Big data: The next frontier for innovation, competition, and productivity*. McKinsey Global Institute. https://www.mckinsey.com/mgi

MIT Technology Review. (2015). *New Boss on construction sites is a drone*. Retrieved from http://www.technologyreview.com/news/540836/new-boss-on-construction-sites-is-a- drone/

Russom, P. (2011). *Big data analytics. TDWI Research, TDWI Best Practices Report*.

Tien, J. (2013). Big data: Unleashing information. *Journal Systems Science and Systems Engineering*, 22, 127–151.

Waller, M. & Fawcett, S. (2013). Data science, predictive analytics, and big data: A revolution that will transform supply chain design and management. *Journal of Business Logistics*, 34, 77–84.

5 Building information modelling for sustainable construction

Introduction

Building information modelling (BIM) is a trend in the construction industry that as rapidly emerged from the shadow into the limelight. It encompasses the integration of modern aided technology in drawing and other constructional activities in the use of computer-aided design (CAD) for designing buildings and other activities associated with it. It is moving away from the traditional method of doing things to a significantly faster way (Lee, 2008). Building information modelling can manage all the information about buildings in a collaborative, digital, and intelligent 3D data environment as long it is properly operated. Project owners have access to all-round information about the building components as it ensures the relationship between these and the corresponding 3D modelling technology. Kibert (2016) and Taylor and Bernstein (2009) further added that not only does it aid in drawing a design that is efficiently analysed, but it also increases productivity in construction. Moreover, the data fused in the BIM process helps in better sustainability of the building. This ensures that it makes a sound improvement to the sustainability of the environment as a whole (Kibert, 2016).

Building information modelling in the construction industry

Kushwaha (2015) stated that the innovation of BIM comprises various models that are generally divided into the following:

- Data management
- Design
- Account of deliverables
- Construction process
- Estimating
- Resource allocation
- Scheduling
- Structural management
- Structural health monitoring
- Supply chain management

DOI: 10.1201/9781003179849-5

Building information models can improve estate values, lessen project schedules, provide accurate costs of estimate, and optimize the maintenance along with management (Eastman et al., 2008). It is also applied in gaining a clear understanding of what the project entails and what the needs are from the perspective of the client or owners:

- For project operation from the perspective of faculty managers,
- For project analysis,
- For project management from the perspective of contractors, and
- Design and further development from the perspective of designer.

(Grilo & Jardim-Goncalves, 2010)

The four primary applications of BIM are the following:

- Integration
- Coordination
- Visualization
- Analysis

(Taylor & Bernstein, 2009)

First, the integration of the 3D view model project embedded in the visualisation of BIM provides a medium for better communication among all the professionals and stakeholders that are connected with such project. It also enhances the process from the planning to the execution of the activities in construction (Taylor & Bernstein, 2009).

A project that is not fully visualised is a hindrance to documentation, understanding, and communication that results in challenges and problems during the construction process. Building information modelling brings to realisation a term referred to as collaboration techniques that enforces concordance among the team members so as to enable for better coordination in both simple and complex projects (Grilo & Jardim-Goncalves, 2010). Coordination in BIM defines and checks disagreement or conflicts faced during the construction phase by reducing rework and avoiding potential problematic issues (Taylor & Bernstein, 2009). All information is stored online for ease of access and usage by the relevant stakeholders (Kushwaha, 2015).

Finally, through the various tools that are integrated into the BIM, the execution and delivery of a project can be witnessed in a very productive way that has not been seen before in terms of cost estimation and scheduling (Khemlani, 2007).

The concept of building information modelling promoting sustainability

Since BIM is the process of bringing together all the persons (client, consultants, contractor, subcontractor, supplier, service engineers, and builders) involved in a construction project to share information in a digital format, the participants in the project form a virtual community of a collaborative hub. There is a single collaborative platform to which all the participants are linked and coordinated. A database of information concerned with the construction, construction expansion, and life cycle issues including details of design, validation is well captured as BIM is analytical and quantifiable in its systematic approach.

In using BIM in the actualisation of construction, the way to sustainable construction projects is not too far-fetched. Building information modelling can promote project sustainability through its use in the building orientation optimization function, a study of the lighting and energy, the analysis of design, and the construction material selection (Azhar, 2011). Holness (2008) mentioned that the contractors, engineers, and designers should work to have a well-coordinated analysis of the building in a coordinated process of design and construction to achieve sustainable buildings that will reduce energy usage and carbon emissions to the atmosphere. Building information modelling is also directed to control the constituent details of the specified design targeted to reduce the energy consumption in building designs and construction. Moreover, BIM implementation through the adjunct of a carbon accounting tracker and weather data contributes to working towards sustainability (Bynum, Issa & Olbina, 2012). Thus, BIM can work towards elevating project sustainability regarding the minimization of water in use and energy consumption, saving of time and costs, resource usage, reduction in waste produced, and material waste, thereby increasing the renewable material used and the productivity at the end of such project (Yang, 2015).

Not only is BIM used for assessment towards performance in terms of analysis of building performance, but it also takes into consideration the impact it has on the social and economic side of the project.

A BIM digital representation of a building would be structured to represent the following:

- The project or building,
- Spaces available within the building,
- Systems present within the spaces,
- Products that make up the spaces highlighted, and
- The relationship or constraints between the spaces marked.

With BIM the structured information behind the models is what really matters because it is what enables the project to be built digitally before being built on site (Cartlidge, 2018). Therefore, in the words of Cartlidge, BIM is a procedure facilitated by software at the higher levels is obtainable by professionals on a construction project, and it is used to enter and store information on a project. The information need does apply not only to the construction phase but also to the operation and maintenance of the completed project. In addition, the software can produce virtual models of the proposed project prior to commencement on site and this facilitates:

- Avoidance of design conflicts,
- Assessment of the environment,

- Essential information on running and maintenance regimes for the client,
- Project planning and resources allocation,
- Consistency in standard of work,
- Sharing of information in real time among the team,
- More precise scheduling of information,
- Links to facilities management systems,
- More exact tendering processes,
- The application of alternative solutions including cost estimation,
- More efficient construction phasing and scheduling,
- Swifter and easier design revisions, and
- Generation of estimated quantities.

Building information modelling for sustainable construction

BIM in design stage

Before the adoption of BIM techniques and innovations, the design stage experiences many setbacks under the traditional procurement route. The design team worked independently of the construction team and design decisions had implications on cost, quality, and time. With the adoption of BIM, however, greater transparency has been achieved during the design phase. Building information modelling has cultivated a more transparent process via the single collaborative platform to which all parties working on the project are linked. With BIM, when design data is added to a shared model, all the involved parties (client, consultants, clients, subcontractors, suppliers, and services engineers) to the project are able to access a practical summary of the products and materials being suggested. In addition to knowing the products and materials, they will know how they will be installed as well as the performance expected after completion of the project.

Efficiency at design and construction phases

Workflow can be reviewed and improved on at every stage because of the 3D level and simulation that BIM technique and innovation provide. Real-time collaboration and orderliness at every step of the process is achieved because the digital representation captures not only the structure but also the spaces within the structure, systems within the spaces, products that make up the spaces, and the relationship/constraints between the spaces. At this combination of phases, many resultant errors from assumptions on hidden details have been removed by the simulations and possible workflows. Clients' satisfaction is largely enabled, and cost and time overrun minimized. The application of best construction techniques and practices ensures that the safety of workmen is guaranteed as well. The collaborative approach has also removed or minimized disputes. Therefore, through the use of BIM for sustainable construction, economic, social, and environmental sustainability is made possible.

Efficient control at operation phase

All the data regarding replacements, refurbishments, and renewals have been captured at the design phase so they are readily available when needed and the smooth running of the completed project and its attaining facilities is guaranteed. Operational recommendations and maintenance schedules can be included to keep a check on enhancing the lifespan of the facilities.

Practical application(s) of BIM techniques as innovation for sustainability achievement

More often than not, most concepts and innovations discussed have been in theory. However, this concept of sustainable construction uses BIM as design; construction and management tools have found application in real life. The Miami Science Museum (MiaSci) is a practical example of a sustainable construction with BIM. The concept of sustainability was incorporated at the three key phases (design, construction, and operation). Its design was modelled to give readings on solar strategies and water systems, while the structure of the building and its orientation impacted easy ventilation. Highlighted components maximized its potential to lessen energy and external resource requirements. Secondly, with orientation MiaSci carefully takes the location of the sun into consideration when designing for the type of project to be executed. The shape of the building is also considered and determined by modelling several angles by considering the position of the sun.

Benefits of building information modelling to the construction industry

Capture reality of the actual construction

With the aerial view of the project to be executed already in vision, the all-round execution of the reality presented through the medium integrated by BIM makes the actualization of the project more understandable. The various parts and components in the device that function differently in the capturing of images, bringing reality through the medium and also helping in getting value for the conceptualization of integrating lasers and mappings obtained from earth for the inputs to give a better and accurate modelling. Digital elevations and other advantages give the user an extra spice that the drawing and modelling on paper cannot offer.

Reduction in the level of waste experienced on site

The integration of the use of BIM not only helps to save time and energy, it also helps to combine activities that could have slowed the construction activities down. The alteration that could be performed/the adjustment that could be made on the software makes it a vital tool to have in some sections of the work, especially when alteration is to be carried out on the initial size of the doors and windows. From this, material wastage is controlled which in turn saves cost and prevents any form of cost overrun or under-costing of the project because of the automatically updated quantities that will reflect in the project work.

Maintain control

It is modelled into a digital system that allows for automatic save to ensure that data recorded and inputted during operation is not lost. The history feature also inspires better understanding in keeping records of operations recorded.

The history portal which can be accessed any time prevents the disappearances of files or a situation where the files will be corrupted, and not be accessible again.

Improve collaboration

Working with the models available through the platform improves the work as collaboration among the teams is well effected rather than working in traditional ways. The speed at which the professionals disseminate information and share any information about the project at any stage of the construction is always a big plus for a construction practice. Since the majority of this added project-management functionality is stored in the Cloud for safety, examples like Autodesk's BIM 360 solutions, makes computer easier and swifter that will in turn increase productivity.

Simulate and visualize

Another advantage of using BIM is the increasing number of tools that allow for simulation and also enable designers to have a clearer view of phenomena such as the sunlight during different seasons or to calculate the building performance in terms of energy, disregarding the hindrances encountered prior to the implementation of BIM. The smartness of the software in applying various rules that are based on physics (of any kind) enables collaboration among the project managers, engineers, and other project team members.

Resolve conflict

The BIM toolset assists in detecting and pointing suggestions to issues that might occur from running the conduit and ductwork in the beam of a supposed construction. By doing the modelling of the project very early in the planning stage, clashes are diagnosed quickly and early, and that could prove to be a very big cost reduction and minimisation of materials to be used for the project. The model could also help in other works that are similar and are conducted simultaneously during the course of construction.

Dive into details

With the intelligence and knowledge transferred during the project execution, major sections of the work to be done are well explained in detail which in turn aids easy communication and further motivates the teams to work better. Through the automation system embedded in the model, quick accessibility to details that might be hard to recognize is analysed and catered for before it is too late.

Take it with you

The integration of a database that could be easily moved (stored) gives more reasons why BIM should be fully implemented in construction to cater for all the sections of work to be carried out. This integrated database gives the user a great deal of opportunity to have range of intelligence at the shortest possible distance in any circumstances. Combining this system with the cloud and Autodesk BIM 360 Build software will further give the user access to the model and project details from any location and on any device with the simplest of information at hand.

Reduce fragmentation

In the days prior to the introduction of BIM, having access to what the world looks like in various forms was very difficult and practically impossible with several unconnected documents in play. Most times took ages for the design teams to see the result of the complex data that are present out there. But combining all a project's documents into a single view through designated folders, BIM enables teams to work without a hitch and communicate seamlessly which therefore leaves more room for effectiveness, thereby increasing productivity.

The summary of the benefits of implementing BIM into construction is represented in Figure 5.1. Such benefits include ability to maintain control of construction projects, reduction of fragmentation, and improvement of collaboration among others.

Summary of the challenges of BIM in sustainable building design

1 Cost of training professionals on the use of BIM software and the high cost of purchasing the BIM software. This was mentioned in the research of Ilozor and Kelly (2012) and Eadie et al. (2013).
2 High level of adaption to the old method of doing things, thereby resisting new change as opined by Ashcraft (2009).
3 Potential legal issues that may arise from the adoption of the new method if not properly utilized (Chien, Wu & Huang, 2014).
4 Increasing complexity in client demands of either new or old technology is another major challenge facing the adoption of BIM (Azhar, 2011).

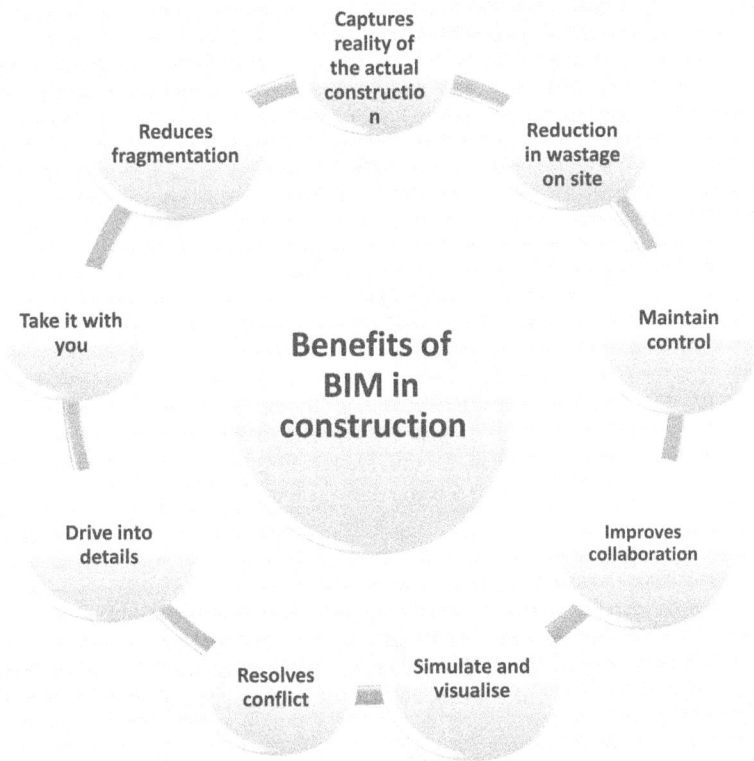

Figure 5.1 Benefits of BIM to the construction industry.

5 Lack of formal understanding about the use and importance of BIM (Arayici, 2011).
6 Lack of ability of the software to work with or use the parts or equipment of another system (Kiviniemi et al., 2008; Eastman, 2011).
7 Increase and hike in work upfront and creativity level (Golparvar-Fard et al., 2013).
8 Unavailability of standard format of BIM contract documents (Gu et al., 2009).

Conclusion

In order to match the rapid progress of other industries out there, the construction industry has implemented technologies and advances that can help in producing a project that is well standardized to meet the demand of the client and the satisfaction of the professionals involved. The involvement in adoption of BIM has instigated an aspect of the work that has not been seen before, giving rise to constructions that are well planned and prepared for even from the contract negotiation stage to the stage when the work commences. The professionals involved can now have access to a wide range of information through the platform presented through BIM. Not only will errors be minimised, absolute data would also be collected to help detect the little faults that may arise later.

References

Arayici, Y. (2011). Technology adoption in the BIM implementation for lean architectural practice. *Automation in Construction*, 20(2), 189–195.

Ashcraft, H. (2009). *Building information modeling: A framework for collaboration*. London: Society of Construction Law. Retrieved from: http://www.scl.org.uk/files/101-ashcraft.pdf

Azhar, S. (2011). Building information modeling (BIM): Trends, benefits, risks, and challenges for the AEC industry. *Leadership and Management in Engineering*, 11(3), 241–252.

Bynum, P., Issa, R. R. & Olbina, S. (2012). Building information modeling in support of sustainable design and construction. *Journal of Construction Engineering and Management*, 139(1), 24–34.

Cartlidge, D. (2018). New aspects of quantity surveying practice. In Cartlidge, D. (Ed.), *New aspects of quantity surveying practice*. London: Routledge, 127–135.

Chien, K. F., Wu, Z. H. & Huang, S. C. (2014). Identifying and assessing critical risk factors for BIM projects: Empirical study. *Automation in Construction*, 45, 1–15.

Eadie, R., Browne, M., Odeyinka, H., McKeown, C. & McNiff, S. (2013). BIM implementation throughout the UK construction project lifecycle: An analysis. *Automation in Construction*, 36, 145–151.

Eastman, C., Teicholz, P., Sacks, R. & Liston, K. (2008). *BIM handbook: A guide to building information modeling for owners, managers, designers, engineers, and contractors*. New York: Wiley.

Golparvar-Fard, M., Tang, P., Cho, Y. & Siddiqui, M. (2013). Grand challenges in data and information visualization for the architecture, engineering, construction and facility management industries. *Computing in Civil Engineering – Proceedings of the 2013 ASCE International Workshop on Computing in Civil Engineering* (pp. 849–856).

Grilo, A. & Jardim-Goncalves, R. (2010). Value proposition on interoperability of BIM and collaborative working environments. *Automation in Construction*, 19, 522–530. http://dx.doi.org/10.1016/j.autcon.2009.11.003

Gu, N., Singh, V., Taylor, C., London, K. & Brankovic, L. (2009). *BIM adoption: Expectations across disciplines*. In: Underwood, J. & Isikdag, U. (Eds.), *Handbook of research on building information modeling and construction informatics: Concepts and technologies*. Hershey, PA: Information Science Reference, 501–520.

Holness, G. V. R. (2008). BIM gaining momentum. *ASHRAE Journal*, 50(6), 28–40.

Ilozor, B. D. & Kelly, D. J. (2012). Building information modelling and integrated project delivery in the commercial construction industry: A conceptual study. *Journal of Engineering, Project, and Production Management*, 2(1), 23–36. DOI:10.32738/JEPPM.201201.0004

Khemlani, L., (2007). Transitioning to BIM. Retrieved October 29, 2012. http://www.autodesk.com/revit

Kibert, C. J. (2016). *Sustainable construction: Green building design and delivery*. Oxford: John Wiley and Sons.

Kiviniemi, A., Tarandi, V., Karlshoj, J., Bell, H. & Karud, O. J. (2008). *Review of the development and implementation of IFC compatible BIM.* Retrieved from: www.eracobuild.eu/fileadmin/documents/Erabuild_BIM_Final_Report_January_2008.pdf

Kushwaha, V. (2015). Contribution of building information modeling (BIM) to solve problems in architecture, engineering and construction (AEC) industry and addressing barriers to implementation of BIM. *International Research Journal of Engineering and Technology (IRJET)*, 3(1), 100–105.

Lee, D. (2008). *Identifying sustainability priorities and engaging stakeholders – The Hong Kong housing Authority's challenges and experience.* Hong Kong: Hong Kong Housing Authority.

Taylor, J. & Bernstein, P. (2009). Paradigm trajectories of building information modelling practice in project networks. *ASCE Journal of Management in Engineering*, 25(2), 69–76. https://doi.org/10.1061

6 Biomimicry and sustainable construction

Introduction

The ways in which people view the world is different. The way we see it and the actions we take are not influenced by the terrestrial environment. Focusing on the study of biomimicry has led some industries to design products and systems that work towards sustainability. The impact of biomimicry has not gone unnoticed. Clough, Olson and Niederhauser (2000) saw the introduction of biomimicry as the link between the interrelationship of the various disciplines that are concerned with technology and science.

The word biomimicry originates from Greek bios (life) and mimesis (imitation) that explains the nature in its relationship with other components along with its corresponding influence in time. There is always a chance to further imitate other products of nature and other materials that are easily available in nature. Even though there are many processes that are concerned with biomimicry, it relates to copying what has been learnt from nature and copied in various forms and processes. We learn and copy from ecosystems that have been tried and tested over time (Zari, 2007).

Some technical challenges can be resolved by the application of the principles of biomimicry on any scale, taking into consideration challenges emanating from other aspects of the work. The advent of biological studies has explained the world in its natural state. The features of some animals and their characteristics are replicated by modern and current trends of technology to be used in various fields. This has helped in translating strategies that are biologically informed into design innovations that are the current trend.

Several authors have mentioned the concept of biomimicry; however, the study by Benyus (1997) suggested that modern and useful information could be developed from considering the beneficial side of embracing those models that are replicated in the application of biomimicry. These models are expressed in terms of the following:

1 Nature as a model: It is helpful to study the models of nature, then imitate or draw inferences (inspiration) from their characteristics in solving human problems.

DOI: 10.1201/9781003179849-6

2 Nature as measure: It proposes the use of standards of ecology to evaluate the 'rightness' of innovations or implementations. Through the billion years of evolution, nature itself has adjusted to what really matters and values over time.

3 Nature as a mentor: Biomimicry is a contemporary way of considering and appreciating nature. It relates a particular time that is based on not only happenings that occur in the natural world but also the events at hand and the information we can learn and obtain from it. Benyus (1997) also stresses the following nine laws of nature, arguing that each identified property should be considered for any truly biomimetic design:

- Nature taps the power of limits,
- Nature recycles everything,
- Nature demands local expertise,
- Nature runs on sunlight,
- Nature banks on diversity,
- Nature curbs excess from within,
- Nature fits form to functions,
- Nature rewards cooperation, and
- Nature uses only the energy it needs.

In nature, there is always a limit to the resource or energy available because organisms multiply until a resource is exhausted. Because of this limit, nature stabilizes the immediate environment and supports whatever function they perform within their ecosystem.

The industrial sector has rapidly embraced biomimicry, which has led to innovations in different fields. However, the challenges of sustainable development have not necessarily always been taken into account (Nachtigall, 2003). Biomimetic is defined as translating good design from nature into design technology. As such it has arrived at a stage where its acceptance as an innovative method is no longer questioned. Beyond technical innovation, looking at principles from nature provides us with insight into deep principles governing life and cohabitation on the planet. In architecture, there are many examples that are influenced by and learned from the natural world.

There are several propagations to architectural design as it seen in network configurations, constructions like those of tree branches of a tree, analogies of flowers and other propagations relating to thinking since the old times, and fusing them into the present. This can be expressed in the following two ways:

1 To transfer the process of emergence of a living entity (like material, form, and structure) to design thinking and

2 To reproduce the form with the concern of finding a befitting design to be employed in designs.

The first concern is of form finding that mostly does not refer to a functional and ecological approach. The second way is a completely different from the first approach that offers the functionality within the nature identified.

Approach and levels

Approaches to biomimicry as a design process typically fall into two categories:

1 This category identifies various needs of humans or defines a design problem and thinks of the possible ways to resolving such identified challenges by considering ways by which other organisms find solutions to similar issues. This is termed *design looking to biology*.
2 Another approach considers some identified characteristics of a particular group of organisms or ecosystems in terms of their characteristics and inferences by fusing them into human designs. This act is referred to as *biology influencing design*.

(Biomimicry Guild, 2007)

The influence of biomimicry in the discipline of architecture has helped professionals to create designs that are relatable. This is a path that leads to a more sustainable environment and buildings. The professional thus relates the concept (level) by considering some attributes or factors in ecosystem, organisms, and behaviour when making projects that are more reliable and environmentally efficient. The organism level talks about how plants or animals may be involved in doing things that are similar from another organism or by copying the whole operations and behaviour of the host organism. The second level refers to the aftermath of the mimicking behaviour exhibited in the first level, by elaborating on how the organism actually relates in a more advanced context. The third and final level has to do with copying every aspect of the system and developing inferences from the observations arrived at which will lead to a path that channels to successful functionality.

Biomimetic design considerations

If we look closer, animals, plants, and microbes are consummate designers. Nature takes distinct approaches for coping with the environment. We take a look at structure built by animals for biomimetic design consideration by learning from their strategies. It is a common practice for animals to always seek shelter or other ways by which they can be safe by staying in hollow lands or underground, hiding in trees, and even staying in isolated caves. Some engage in parasitic arrangements while others have other modes of habitation. We also know that some animals and birds build their own nests, shelters, and other features where they could live to mate or continue the ecosystem by interacting with each other.

Structures built by animal, or animal architecture (Hansell, 2005), are bound with nature, unlike those that are human-made. Animals create their constructions with sophisticated features that allow them to survive. These include traps, temperature regulation, special purpose chambers, ventilation, bait, multiple escape routes, structural strength, and many other features. Animals also build their constructions with a limited energy and within an ecosystem. For example, termites' mounds are perfect natural constructions, with an efficient passive ventilation system that can maintain a stable interior temperature regardless of the exterior temperature. Termites make their mound from wasted materials of plants and animals around their local area. Their construction process produces nitrogen, phosphorus, and organic materials that help to enrich the soil, fostering more plant and animal growth in the area (Turner & Soar, 2008).

This is a perfect example that shows that apart from the termite mound construction being efficient, the process of their construction also has a positive impact on their environment. This is an important lesson for us as architects to learn and improve on our construction design process and industry.

In this study, we have chosen the design and construction of birds' nests and birds' behaviours as nature's role model. The constructive behaviour of birds thus seems likely to indicate potential paths towards a sustainable architecture.

Nature's role model: The constructive process in birds and their criteria for sustainable architecture

The constructions of birds are durable. They have been modified and perfected by the process of natural selection and only the most suitable nests have allowed the species to reproduce until today (Mainwaring et al., 2014). Given the rise of biomimicry, their study seems legitimate to indicate possible paths towards a sustainable architecture. This study demonstrates the design and the functions of the bird's nest along with the way they have influenced the construction disciplines.

The architecture of birds shares several features with that of man. They both make a distinction between an interior and an exterior, often by assembling materials into a coherent structure. They follow the same principles of solidity, utility, and even appearance. The nest-like the house must be solid, meet a need, and give a specific image, as in the case of birds whose construction is camouflaged in their environment so as not to alert the predators. This therefore points to the implementation of a particular technique with particular materials, on a particular site.

Chronologically, the first concerns the selection of a locality in which a type of nest is made of specific materials. The construction can begin a second time; effective constructive techniques are adopted because they meet the constraints related to the site and the morphology of the constructor. The cycle is completed by the exploitation of the built nest and the end of life. We extract the strategies of how birds design their nests, how they construct, and how they operate with limited energy and sustainability in comparison with the way we design, construct, and operate our buildings.

Birds are the most persistent inventive builders, and their nests set the standard for functional design in nature. Birds build some astonishing structures, from nests the size of walnuts to makeshift rafts and even apartment complexes (Mainwaring et al., 2014). The bird chooses the nest site with the utmost care, for the reasons of safety, accessible construction, and suitable local materials for transportation.

Biomimetic design methodology

Biomimetic is defined as the process of formulating general concepts by abstracting common properties of instances of good design from observations in nature (Vincent et al., 2006).

The approach of biomimetic design methodology is basically a three-step process:

Research → Abstraction → Implementation.

(Nachtigall, 2010)

The research concerns the selection of nature's role models to suit a specific problem in design, while the abstraction concerns the analysis of nature's strategies transferred into the design phase. Implementation concerns the ability to build the biomimetic design concept according to construction criteria.

Biomimicry to improve sustainability

Biomimicry in built environment is related to social issues and also environmentally based challenges such as unsustainable waste production, the amount of energy used and resources allocated for use, as well as greenhouse gas emissions related to human habitats. A strategy to reduce these harmful effects should be implemented with this constant development of urban buildings. Biomimicry proposes a creative and environmentally tenable systematic approach that can provide solutions that are compatible and permeable. The conservative nature of the environment is transposed into a situation whereby nature itself will adapt to conserving natural phenomena that make up the whole cycle of survival. The beauty of nature and the characteristic adaptation to managing resources and materials to avoid wastage are essential for the long proposed way of managing the environment; from the way the sun radiates solar energy to various absorptions through different mediums as well as the utilization of these energies in different ways to maintain and sustain life.

The application of biomimicry enables an avenue for better and efficient buildings by offering a better understanding of the precepts of the type of building to be erected even before it starts and after it has been completed. The conservation of the energies, materials, and other elements is vital in working towards an environment that could be sustained for a very long duration. Nature is a great example to live in harmony with. For instance, plants use air pollution from the environment and turn carbon dioxide into oxygen that humans inhale. Considering the level of biomimicry as it affects the all-round functionality of the environment, and mimicking an organism on its own without replicating how it participates in the wider relation of stages, its affirmations in the ecosystem in which it is situated can result to making designs (from conclusions) that are not even related to anything observed in the first place. Since imitating a species is only a particular function, for example, constructing a building in the form of cactus (simple copying of the shape), it may not increase the general sustainability of construction. Biomimicry at the action level mimics the organism's actions. In biomimicry at the behavioural level, the organism's (action) in mimicking or behaviour when used in combination with other rates of biomimicry (organ's) method can also be used in various temporal and spatial scales. At this stage, designers need to decide whether the conduct of the organism is appropriate for humans to emulate and which part of their behaviour can improve sustainability building. Nevertheless, it would not be acceptable to emulate the social structure of termite colonies if universal human rights were not considered. Biomimicry at the environment level has the advantage that it can be used on two metaphorical and functional scales. Putting the high-rise buildings into the conceptual framework of construction helps to maximize the available of land by reducing the amount of floor area used. The land area is used effectively rather than having small buildings that occupy floor area that could have been used for another construction. The construction of large buildings is now being embarked upon because of the limited supply of land. Another factor that gives rise to large buildings is the rapid growth of the urban amenities and population that seems not to be dwindling.

Tall buildings with an average height of 50–300 m are divided into three forms.

- Tall buildings that form of building makes up 90% of the world's total high-rise buildings.
- Super tall buildings of 300–600 m of average height make up 10% of the total high-rise buildings.
- Mega tall buildings of 5–300 m of average height represent 0.05% of the world's total high buildings.

Living in high-rise buildings has its own benefits. It can provide housing people near their workplace, resulting in fewer work trips and lower fuel consumption. The new high-rise buildings now have social interaction facilities such as entertainment centres, parks, workshops, libraries, gyms, and public spaces so that people do not need to go looking for such amenities once they are at the summit of these buildings. Upper floor residents often enjoy the pleasant aerial view of the surroundings even though it may be difficult and sometimes troublesome to live adjacent to many strange neighbours.

Conclusion

Considering every aspect of sustaining the environment as well as the components that comprise it is synonymous with working towards concepts that are channelled in the direction possible. Implementing the necessary elements that correspond with the measures on the ground to combining biomimicry and how it affects construction practices will take the disciplines involved to new height in terms of cost, durational usage, and resources management. The study of biomimicry will enhance the construction practice and bring to light new aspect of the work that may not have been considered before. This chapter highlights the interrelationship between the living elements of the environment and the effect they have on plans drafted towards working towards a sustainable culture and practices.

References

Benyus, J. M. (1997). *Biomimicry: Innovation inspired by nature*. New York: Morrow.

Biomimicry Guild. (2007). *Innovation inspired by nature work book*. Biomimicry Guild.

Clough, P. M., Olson, K. J. & Niederhauser D. (2000). *The nature of technology: Implications for learning and teaching*. Springer, United Kingdom.

Hansell, M. H., (2005). Animal architecture. New York, Oxford University Press, England.

Nachtigall, W. (2003). *Bau-bionik: natur-analogien-technik*. Berlin: Springer Auflage.

Nachtigall, W. (2010). *Bionik als wissenschaft: erkennen-abstrahieren-umsetzen*. Berlin: Springer Auflage.

Turner, J. & Soar, R. (2008). Beyond biomimicry: What termites can tell us about realizing the living building? *First International Conference on Industrialized, Intelligent Construction (I3CON)*. Loughborough University.

Vincent, J. F., Bogatyreva, O. A., Bogatyrev, N. R., Bowyer, A. & Pahl, A. K. (2006). Biomimetics: Its practice and theory. *Journal of the Royal Society Interface*, 3(9), 471–482.

Zari, M. P. (2007). *Biomimetic approaches to architectural design for increased sustainability*. Auckland, NZ: Woodhead Publishing.

7 Blockchain for sustainable construction

Introduction

The major challenge hampering the involvement of the construction industry in modern trends is its stalling view towards embracing technological advancements, comparing it with breakthroughs that have been experienced in mechanical engineering industries, automotive, and logistics (Mason & Escott, 2018). Any activity that is related to construction is a framework of several participants, concepts, planning, designs, materials, products, processes, and other frameworks not highlighted. Monetary activities are frequently performed with the progression of the project. With disputes arising from possible mismanagement of funds allocated, several measures have been employed to solve either old or just occurring crises at every phase of construction. Although arrangements such as terms of payment, conditions, and confidentiality of the agreed data are spelt out in a contract or an agreement, disagreements often occur from conditions that might have been overlooked in the stipulations of the framework that had been agreed upon before (Cardeira, 2015).

Blockchain is a profound digital ledger of economic transactions that can be programmed to record not just financial transactions but almost everything of value. In blockchain, it takes just one party to initiate it. The party initiates payment by means of a block, and this block is verified by thousands or millions of computers (as new data is uploaded and later stored to create a unique one that it stores automatically and records with a unique history that is done in an automated way). When a single record is tampered with, it affects the whole record making blockchain a trustworthy innovation.

Blockchain technology was invented by Stuart Haber and W. Scott Stornetta in 1991 with the aim of securing document time stamps so as to avoid hacking. Blockchain is part of the models generated by the digital world to revolutionise the digital world to a further view of security resiliency and efficiency of the system. In making transaction using blockchain, the following analogy explains how it works between the participants involved:

- There is a want to send money from one party (A) to another party (B),
- The transaction is represented online as a block,
- The block is broadcasted to every party available within the network,

DOI: 10.1201/9781003179849-7

- Approval of the block by the operators of the network,
- The block is then added to the chain which provides an inerasable and transparent record of transaction, and
- The money is moved from party (A) to party (B).

(Wild, Arnold & Stafford, 2015)

Chance LLP, one of the world's pre-eminent law firms in a brief report after looking into the Blockchain Technology in 2017, referred to it as integration of the following elements that are explained below.

Distributed ledgers

Blockchain is a type of distributed ledger technology as this is a basis that is copied all around many places as a ledger, a history of transaction. Each of these ledgers is synchronised across the network of the system. This helps transactions to have public witnesses, thereby making fraudulent activities mostly impossible.

Public key cryptography

Lord (2018) explains that this uses two pairs of digital keys, namely a private key and a public key to code and decode data so as to prevent it from unauthorised parties. If a party wants to code a message, the intended recipient's key is obtained from the public contact when the message arrives; the receiver then decodes it by using a private key to which only the party has access.

Consensus mechanisms

This is a tedious way of making sure only valid data is being added to the blockchain (Clifford Chance, 2017). In a blockchain, by calling for transactions to be definite with rules, all computers agree on the same decision made. Once the transaction has been done, the updated transaction is updated on the blockchain with the new entry automatically being available immediately in every aspect of the system so that all of the computers end up with a similar recognition of the ledger.

Miners

This is where the question is raised about who is responsible for choosing transactions that go into each block is tackled. These are those who digitally sign the blocks created.

Pillars of blockchain technology

Blockchain runs through some pillars; these pillars make it unique across its operations and output. These pillars are identified as decentralization, reliability, authentication, immutability, and transparency as shown in Figure 7.1.

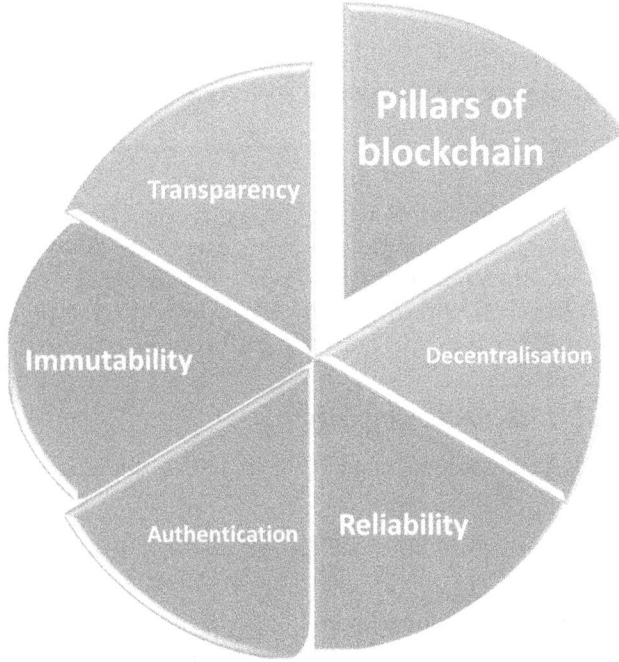

Figure 7.1 Pillars of blockchain technology.

Decentralization

This includes operation across a system of networks made up of computers. Where transparency is high and parties can be compensated for outcomes, projects can benefit from a more decentralised and agile method of approach as well as for work performance even though projects are well structured and contract-based with clear objectives of meeting the time, specifications, and rework.

Reliability

All these nodes now have the same replica of the blockchain that is put through a process and helps highlight any irregularities.

Authentication

Turk and Klinic (2017) stated that in the bitcoin blockchain, there is a system that is called the proof of work mechanism that is engaged to confirm transactions. This is characterised by the blockchain of bitcoin. It employs a scientific and exchange issuance mechanism to reward its miners.

Immutability

Once blocks are chained, there is no going back. Blockchain offers that in its sequence and that guarantees a safe network in which its operations are embedded.

Transparency

There is no central authority; information is available for everyone to see. There is an access password to log into the chain but it cannot be easily hacked because there is digital information stored in a public database and hacking will be detected easily. In public blockchain, it is barely possible to change a block owning to the transparency that is spread across the network.

Challenges of blockchain technology

The challenges facing the adoption of blockchain technology are discussed below;

Technical challenges

Owing to the blockchain technology still being in the nursery stage of development, it encounters challenges such as the long time it takes for transactions to be affirmed as well as the long time that is still needed to download information.

Human-related challenges

Inadequacy in the awareness coupled with basic understanding of how blockchain technology works is an issue to be resolved before there could be massive acceptance and adoption by the proposed users in uploading their personal information in a decentralised manner as privacy is always considered in situations like this. Owing to technical and regulatory uncertainties, most corporations are still in the exploratory phase; therefore, the awareness expected to have generated is a little on the low side.

Problems blockchain solves

The implementation of blockchain technology solves various challenges in construction. These are discussed in subsequent sections.

Payments

There is the issue of timely payments after the awarding of certificates to stakeholders such as project bank accounts, bonds, defects, and retention trust. Blockchain helps to improve the accountability and transparency of the payment process in order to enable secure and traceable payments. It prevents incorrect information by minimising human errors.

Work health and safety management systems

Blockchain helps to avoid manual logging of site events and also provides a secure origin of workers' health and safety, including information that is concerned with the site, with a greater sense of accountability in systems and management.

Smart contracts

Blockchain helps to reduce paper shuffling between intermediaries as it involves payments, intellectual property rights, materials, and equipment. These call for automated, tracked, and transfer of ownership during the onset of the project.

Effect carbon tracking

Much distributed or transferred information is stored and transferred onto a common platform that helps to track both embedded carbon and operational carbon that is an increasing challenge for construction.

Benefits of blockchain technology

Near-real time

In determining the best time to make settlements, blockchain helps in enabling a near flawless time by recording the transactions embarked upon which therefore nullifies potential conflicts and systematically reduces risk.

Direct transactions

Blockchain technology is established on digital signature that allows any two parties to be involved in transactions directly with each other without the need to involve a third party or any form of regulatory or government authority.

Distributed ledger

The distributed system network records the history of all the transactions carried out, small or large. Blockchain stores all the transactions, whether small or large, through its dedicated platform and can further furnish proof that such trade actually happened. This also prevents any activity related to fraud or other inappropriate dealings.

Other benefits of blockchain

1 Since there is no involvement of a third party, there is no cost incurred by employing a third party and this eliminates complications.
2 It helps to maintain data and information without any organisation or government interference as transaction is done in a controlled system without a central owner.

Sustainable procurement in the construction industry enabled by blockchain

The construction industry procurement processes and the implementation of the sources of materials can be improved by another essential aspect of the whole supply chain, namely sustainability. Tata Steel conducted a survey about the possible ways in which sustainability is profitable to all and the principles of responsible sourcing in the industry. It found that the background of materials was particularly crucial as the reuse and reprocessing of materials depends on the certified material specifications.

However, impactful sustainability is often driven by considering the whole lifecycle of a structure. This includes the design, construction, its sustainable procurement, operation, and maintenance through to demolition. 'Imagine payments for labour and materials being triggered automatically throughout the supply chain as progress is made with zero financial reconciliation required. Thus a client knows the material originator will receive payment for a specific project, ensuring both fairness and completeness of warranty', commented Tim Rook – Associate Partner, IBM blockchain and digital transformation. Currently, it can be difficult to confirm the specification and source of the methods, materials, and other activities involved in the building after construction has been completed. It can also support the planning of site waste management that usually relies on data from the supply chain (e.g. invoices, delivery note, and material specifications).

Fostering enhanced efficiency and trust in the supply Chain

Clearly the prospects of utilizing blockchain technology are blossoming regarding the background of applied materials, and how a construction project is procured. The system, with a see-through chain of custody, would inspire good behaviour and showcase all forms of quality throughout the entire supply chain. For example, if structural materials fail to meet up with quality standards and criteria, regulators will have grounds to identify the culprit through the blockchain-enabled chain of custody. In this way, honest and trustworthy suppliers can easily be identified and incentivised to maintain quality certificates in order to establish long-term relationships. Such trust can move the supply chain of the industry from one-off transactions to a more integrated, relationship basis.

Blockchain and sustainability

For a sustainable industry, more efficiency and greater productivity are required. Blockchain, recurrently hailed as possibly having a greater impact than the invention of the Internet (Jenkins, 2017), might have the possibility to increase productivity, increase efficiencies, and minimize costs. Accenture reported in 2017 in their value analysis of blockchain for the finance industry that blockchain technology could save the world's largest investment banks, the industry where blockchain was first exploited, up to $12 billion a year that is more than the annual GDP of half the world's countries (Accenture, 2017; Anon, 2018).

Technological development can have positive outcomes for economic, social, and environmental sustainability (Purt, 2011). Blockchain discussions are gathering momentum as a prospective key promoter for unlocking sustainable growth and productivity improvements, and it is a real game changer for construction (Gueguen, 2018). However, there is little simplicity within the industry regarding how to accomplish this task. It is vital for the industry to understand the potential opportunities for better quality business sustainability that may be unlocked in time with this emerging technology, even if the conclusion is not to actively develop blockchain solutions at this time.

One of the applications of the blockchain innovation in construction is smart contracts. A smart contract is essentially a computerised agreement (contract) that can execute its terms automatically when the highlighted conditions are met. It may very well be utilised to expand further procedure effectiveness and traceability. With such a framework, trust and joint effort would be improved in a general sense among all members. Smart contracts are at the forefront of the most exciting platforms as it opens doors for simplicity as introduced by blockchain innovation. Thanks to smart contracts a wide scope of procedures can be raised, automatized, and in the future become increasingly powerful. Let us look at a theoretical example; at a building site each worker who enters the site shows his ID card for security, safety, and health reasons. The data about who entered and the duration of time spent working around the activities on site are captured and listed on a blockchain-empowered distributed ledger between the client, the expert or consultant, and the contractor. Thus, there is no administration involved to approve this data, as it has been programmed on the blockchain.

The following are the advantages of utilising smart contracts according to blockchain innovations in the construction industry:

- **Accuracy:** If contractual conditions by which the contractor has to abide are accurately registered on a smart contract, following such terms and monitoring of conditions are precise. Contractual collaboration that is supported and automatised with smart contracts can significantly reduce the number of claims, conflicts, and disputes in terms of the time spent in solving them, thereby improving stakeholders' relationships.
- **Transparency:** Every payment made, transaction carried out, business interaction done, and execution can be registered on the blockchain, thereby making the whole process transparent and reliable.
- **Risk management:** The collaboration of smart contracts can ensure that activities can well be managed and complex sections of the work can be divided up to enable a better understanding of the project. When risk management has been achieved, risks that might be associated with payment are reduced to the barest minimum.

- **Compliance:** Contractual standards such as NEC3 and NEC4 could be critical and implemented part of smart contracts. Together with the project information accessible on blockchain, regulatory compliance can be demonstrated conveniently.
- **Cost effectiveness:** Noteworthy cost investment funds can become overheads, administration, and project control. Also, project procurement data are signed in a detectable manner for opening project assessment and cost optimization with bits of knowledge.

To overcome cultural barriers, the practices in the construction sector will extend to widespread blockchain adoption if participation (contractors and subcontractors) can be based on relationships that have existed for a long time because the industry is relationship based due to the many privately owned (family) firms and private companies. Construction is advancing technologically in all aspects of its activities.

Opportunities related to blockchain in the construction industry

1 Collaboration is increased owing to the transparency in data sharing that increases trust between parties.
2 Valuable information is obtained by all stakeholders due to the digital twinning of a built asset throughout its whole lifecycle, even up to the development for sustainability. This will help in aspects whereby when a product or a specified detail is needed to be located in a building. Finally, there will be a place to conduct a simple search to acquire the information needed in the building.
3 Removal of all intermediaries and all transactions executed are guaranteed as smart contracts automate the processes and the payments.
4 Faster and more streamlined processes as multiple verifications are removed.
5 Due to the removal of intermediaries, all transaction costs attached to them were removed, paving the way for a low cost.
6 Reduces human error.

Conclusion

Blockchain enables the ease in keeping records of transactions that were considered to be difficult. It bridges the gap between the users and the intermediary by making available, but safe large amount of information within a reasonable limit. With the amount of information present at the different stages of construction, it is necessary to ensure the safety of every detail that is going to be involved from when the project is planned to when it will be delivered. The absence of a third party brings down the cost to be incurred and makes information safer since it will be limited to certain people and a respective line of function.

References

Accenture. (2017). Blockchain technology could reduce investment banks' infrastructure costs by 30 percent. Accenture Report (NYSE:ACH).

Anon. (2018). *What is blockchain technology? A step-by-step guide for beginners*. Retrieved from: https://blockgeeks.com/guides/what-is-Blockchain-technology

Cardeira, H. (2015). Smart contracts and possible applications to the construction industry. In: *Proceedings of the New Perspectives in Construction Law Conference*. Retrieved from: https://heldercardeira.com/1503P.pdf

Clifford Chance. (2017). Blockchain: What it is and why it's important. Retrieved from: https://talkingtech.cliffordchance.com

Gueguen, A. (2018). *Blockchain technology and BIM: Building-in proof*. Retrieved from: www.pbctoday.co.uk/news/bim-news/blockchain-and-bim/45972

Purt, J. (2011). *Name of topic required*. Retrieved from: https://www.theguardian.com/sustainable-business/live-discussion-sustainable-technology-socialenvironmental-change

Jenkins, H. (2017). *Will the blockchain be bigger than the internet?* Retrieved from: https://www.tech.london/news/will-the-Blockchain-be-bigger-than-the-internet

Mason, J. & Escott, H. (2018). Smart contracts in construction: Views and perceptions of stakeholders. In: *Proceedings of FIG Conference*, Istanbul, Turkey. Retrieved from: http://eprints.uwe.ac.uk/35123/

Turk, Z. & Klinic, R. (2017). Potentials of blockchain technology for construction management. *Procedia Engineering*, 196, 638–645.

Wild, J., Arnold, M. & Stafford, P. (2015). Technology: Banks seeks the key to blockchain. *Financial Times*. Retrieved from: https://www.ft.com/content/eb1f8256-7b4b-11e5-a1fe-567b37f80b64?segid=0100320#axzz3qK4rCVQP

8 Cryptocurrency for sustainable construction

Introduction

The earth's potential to maintain its existence has been questioned for so many reasons. Its sustainability has been reduced over the years due to ever-increasing pollution (of various kinds) by man, especially when considering the development of the Industrial Revolution that now forms the basis for the economic development of most countries. If we do not change our activities and if we continue the way we have been living, we will strip the planet of its ecosystems, leaving it unable to sustain life in the very near future. The contemporary form of doing transactions has impacted in many ways towards human survival. The over-exploitation of some reusable herbal sources, the exhaustion of non-reusable assets such as minerals and fossil fuels, and the gases released during the use of these assets give rise to adverse effects on the planet in terms of ozone layer depletion and air and water pollution, among others. Since money is involved in practically all human activities, it is advisable to cater for forms by means of which transactions could be made with regard to the sustainability of the earth, especially that of us (human beings).

The ever-increasing demand and market for cryptocurrencies in the past years has alerted the attention of government officials and concerned parties to consider the significant impact it is having on the economy. Almost every year, the value of cryptocurrencies is on the rise hence the value people place on it. With this information, some regulators are contemplating whether to regulate, and to what possible ways of doing so. Even at that now, there is no unanimous view on the ways by which such implementation could be effected as regarding the safety of cryptocurrencies; factors such as volatility of the price, and problems with hackers have called for more supervisory measures and this has contributed immensely to the difficulty (Houben & Snyers, 2018).

Definitions

Cryptocurrency is defined as a distributed, decentralised digital currency based on cryptographic principles. It is also explained as a digital platform for making transactions by making use of specified cryptograph in securing any transaction,

DOI: 10.1201/9781003179849-8

and make control of extra units applicable and affirming a current state of trans-action. Cryptocurrencies use a non-centralized system that is in contrast to the centralized digital currency operated in the banking sector. Cryptocurrency is also defined as electronic money created from code and systematic interaction between participants.

Bitcoin is a popular example of cryptocurrency. It is a decentralized digital cur-rency, or a unit of this currency. It can also be defined as an electronic currency that transacts without the aid of the central bank or single administrator. It can be used to transact from one party to another on the connected bitcoin network without the need for a middle platform.

Cryptocurrency fulfilment in the role of money

Cryptocurrency has grown in prominence and reputation in society, and it is altering day-to-day lifestyle activities as the world is becoming digitalized to an increased degree in the 4IR. One of the fundamental benefits is the decentrali-zation in cryptocurrency, which enables the verification and safety of money not by means of a large, central, national group, but rather by a peer-to-peer network. Little sense – The theory is that, as with existing fiat money, no state or central bank can directly control the supply. Therefore, cryptocurrency can isolate its value from any potential governmental or political upheaval or the devaluing of certain currencies. It provides its consumers with a sense of security and stabil-ity that conventional, sovereign currency does not. This feeling of security and stability enables cryptocurrency to be widely utilized as a medium of exchange for items and services, and for people to intrinsically retain the value of their money, understanding it is not related to a central institution. Finally, as a unit of account understanding, it can use for more accurate basis for charge and value calculations. Furthermore, a public ledger is maintained with the aid of a public community of miners, alternatively of a large, sovereign bank. This system and absence of an intermediary now not only permits transaction fees to be drasti-cally lowered, but it also transcends borders to enable quicker transactions. Just as savings cards have allowed for extra international transactions that were now not possible with paper money alone, cryptocurrency permits better transaction with quicker delivery within the shortest possible period.

As the 4IR is digitalizing daily activities in human life, cryptocurrency is tak-ing over the role of cash from money or plastic savings cards in transactions. The European Central Bank (2009) has described three functions of money as discussed in the following sections.

Medium of exchange

Cryptocurrencies can serve as the medium of exchange between parties that are involved in a transaction through whatever means possible. It is however to be noticed that acceptability is dependent on individual's perspectives and how much it can stand the test of time. The use of cryptocurrencies has somehow reduced the over-reliability on money as the major way of transaction. It is serving as a medium by means of which deals could be completed within the shortest period of time. Another advantage is the flexibility of the medium as the user may decide to pick any medium available rather being restricted to just one particular way. Taking Bitpay as an example, they are the largest bitcoin fee processors and they declared that over 100,000 transactions were carried out in 2015. Since this company owns 50% of the price market in bitcoin, it is then safe to say that an assumed 200,000 transactions were carried out and completed if we are to go by the volume. This automatically sets the value of bitcoin as a well-known form of medium of exchange.

Store of value

There has been much disagreement over the value that bitcoin brings into a transaction, taking into consideration how the transaction is done with no specified regulations. However, this is to be disputed because the value that a local currency has is what bitcoin currencies also carries. The difference lies in the way it is being carried out, namely digitally as the case is in cryptocurrencies rather than the traditional way of spending money. The value it carries is also established as it can be exchanged with other currencies. In addition, the process of making it available is through the 'digital wallets' since it cannot be deposited in a bank. This has definitely protected from hacking and other fraudulent manipulations that might occur regarding currencies in banks.

Unit of account

As a currency, the ability of cryptocurrency to meet the three requirements of the exchange medium, account unit, and value store has been discussed briefly in the introductory part of the subject. Notwithstanding, the views shared in terms of relevance and sustainability are explained by Carrick (2016) that asserted more positive outlook on the role of cryptocurrency as a medium of exchange. Bitcoin is used for transactions, and countrywide backing should no longer be too strong a criterion for backing a currency, i.e. susceptible currencies of Cambodia, Laos, and Uganda. It raises necessary issues regarding the function of cryptocurrency in satisfying the characteristics of a medium of exchange.

For a unit of account, Sauer (2016) argued cryptocurrency satisfies this feature for those who take delivery of cryptocurrency. Hence as cryptocurrency grows in utilization and popularity, it can effortlessly fulfil the function of unit of account. Carrick (2016) also argued that bitcoin can be divided into a countless number of pieces. Bitcoin is fungible. All bitcoins are created equally, and they all can be interchanged. Bitcoin is countable and poses challenge to mathematical operations that is set to be finite. It implies that even if there is a finite set amount of cryptocurrency, it can nevertheless be broken down into components divided similarly to fulfil the function as a unit of account.

Lastly, money can be said to have similar characteristics when value is mentioned as shop account. Sauer (2016) and Heller (2017) argued that the fee of cryptocurrency, at present, is not stable and the shop of cost now is not met due to the rate volatility. However, Carrick (2016) contended that as a shop fee of account, it can be used to diversify chance and in fact, created to forestall misplaced due to governmental action. Hence it claims that non-reliance on governments would hinder the value of cryptocurrency due to restrictions in some transactions. If ordinary people who wanted to invest money were given the choice of a country with a questionable regime in power or cryptocurrency with its blockchain technology, most people would choose the latter.

Advantages and disadvantages of cryptocurrencies

Ivashenko (2016) mentioned the benefits and disadvantages of bitcoin as discussed below.

Advantages

1 **Code for mining cryptocurrency**: bitcoin (BTC) applies the similar principles and guidelines that online banking systems use. The major difference is the disclosure of information about the users that does not happen in cryptocurrency. All the data about the transaction in the BTC network is shared by the users. However, there is no information about the receiver or the sender of the coins (there is no gaining admission to the private statistics of the owner's wallet).
2 **No inflation:** The widest variety of coins is strictly restricted by using 21 million bitcoins. Since there are no influences whatsoever from political forces or companies in a certified position to change an order, there will be no nature and evidence of development of inflation in the system because of the standard policy imbibed into it.
3 **No boundaries**: Payments that are transacted into this platform are impossible to nullify. The coins cannot be counterfeited, copied, or spent twice. These skills guarantee the integrity and transparency of the entire system.
4 **Peer-to-peer cryptocurrency network**: Since there is no specific server that is accountable for the transaction tied to a regulation, the expected numeric exchange is between parties through a medium of using software that has been programmed to handle such transactions. The selected installed program-wallets are all sections of the bitcoin network that are linked together. The interested user then looks for files dedicated to transactions and searches through the options available in the bitcoin wallet. The involvement of the government and others is restricted because they have no control over what goes into the system.
5 **Unlimited possibilities of transaction:** As wide as the sea is, so much more are the unlimited transactions that can be carried out using cryptocurrencies. The possibility of using this medium cut across any platform provided that the users are well within the agreed conditions and there is no indication of any hindrance or mistrust at any time. In simpler terms, you can transact whenever you feel like it.

6 **Low BTC operation cost:** The operation hobby costs go to BTC miner's wallets. The benefit of cryptocurrencies also combines that of an e-commerce along with other functions it is used for. Since it is a substitute for cash, payments of any sorts to any financial institutions are greatly reduced to the barest minimum. So, the cost in transacting any deals is reduced to the lowest minimum that is estimated at about 0.1% of the amount transacted. This is low when compared to the monies to be paid to institutions that are involved as a third party between users. The percentage deducted is then transferred directly into the miner's wallet and not any other medium.

7 **Anonymity:** It is completely anonymous as it contains no identification details and at the same time it is absolutely transparent. Business enterprises of any kind can make any number of bitcoin transactions without making any reference to names, addresses, or any other information. This anonymous nature could secure the privacy of the statement of transaction. This anonymity ensures that the transaction is private.

Disadvantages

Ivashenko (2016) listed the dangers as follows:

1 Since there is a great deal of unpredictability of the fee allocated to cryptocurrencies charges as well the swiftness in transaction, governments of countries could be forced to negate its use in policies that could affect economic policies. For this reason, this volatility is always questioned and its total implementation could be far from being endorsed.

2 There is a possibility of the system crashing since it is not regulated by identified agencies that could monitor its operation when users are so many on the platform.

Cryptocurrency for sustainable construction

The construction industry has been mentioned as one that aims to work towards a stable environment through the use of various practices, management strategies to bring out best constructions in executing projects. Examples are the implementation of capital infrastructure projects round the world that have an exceptionally fragmented, scattered and complicated designs, but aesthetic and sustainable at the same time. For example, the Cross rail project in London, with more than 700 various suppliers from the United Kingdom, or the Burj Khalifa, with more than 12,000 workers from more than 100 countries on site at the peak of its construction. To manage such prolonged chain of work in progress, schedule, fee and payments, vast effort, and assets are needed. Owing to these challenges, development initiatives experience distinct forms of errors, accidents, and delays at various levels and to a varying extent.

The inadequacy of accountability in the construction industry (especially large constructions) is an ongoing issue that has crippled the development of

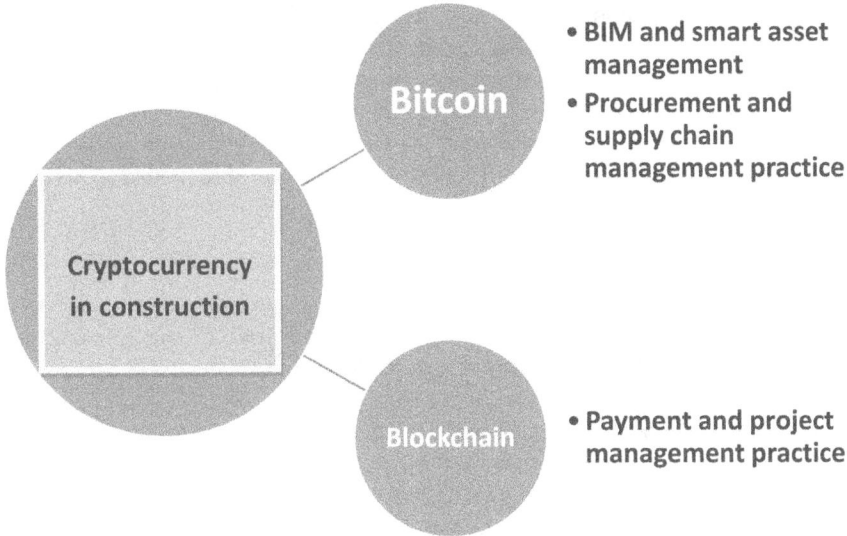

• **BIM and smart asset management**

• **Procurement and supply chain management practice**

• **Payment and project management practice**

Figure 8.1 Application of cryptocurrencies in the construction industry.

the industry for many years. Companies are compelled to find ways to cut corners and deflect blame from the resulting failures. These are accurately the main "pain points" and areas where cryptocurrency can help and make the system more efficient, simpler, and more accountable concerning everyone concerned in the project. There are manageable cryptocurrency functions that have already been introduced and that have had an impact on the economy. Some of them can be applied to the construction industry without delay and some of them can serve as groundwork for a smoother running of capital development projects.

Cryptocurrency can probably be applied in the construction industry in the following areas as shown in Figure 8.1:

• BIM and smart asset management,
• Procurement and supply chain management practice, and
• Payment and project management practice.

Application of cryptocurrency as means of payment

In general, the use of cryptocurrency in an economy will increase the economy cash flow as well as other functions would be made easier to monitor and manage i.e. change of money and security risks are reduced. There is also a reduction in monetary money float crunch, a reduction in disputes occurring as an end result money change. These are also benefits of cryptocurrency. The blockchain technological know-how underpinning cryptocurrencies is exceedingly resistant to counterfeiting that ensures financial security. In addition, it prevents any arbitrary addition to the supply, the double spending of coins and transactions cannot be repudiated or undone. The nation of the blockchain and each change to it are verifiable by way of every user. There is no danger of inflation as it is strictly constrained by means of 21 million bitcoins. As there is no political interference nor are firms able to exchange this order, there is also no possibility of inflation in the system, and no boundaries that also mean that repayments made in this system are impossible to retract. The coins cannot be counterfeited, copied, or spent twice.

Conclusion

The emergence of the usage of several cryptocurrencies in the construction industry and other industries has put up a platform that is safer, more acceptable, and more secured than the traditional ways employed before. Blockchain (see Chapter 6) showed the expected positive effects of cryptocurrencies when it is compared to benefits experienced when it is not applied in the past, and when it is being used in the construction industry and relative ones. It links several competencies that were not possible until a particular stage of the construction has been reached. However, now with the help of cryptocurrencies, much can be done with the time being minimized and efficiency being maximized. Furthermore, other currencies embedded in it can help in terms of working towards a problem-free construction. For example, bitcoins are not affected by inflation that is a major factor in developing countries as it follows the laid down guidelines strictly and that will keep the cash flow of the contract sum constant no matter the economy of the project location.

References

Carrick, J. (2016). Bitcoin as a complement to emerging market currencies. *Emerging Markets Finance and Trade*, 52(10). DOI: 10.1080/1540496X.2016.1193002

European Central Bank. (2009). Annual Report 2009. Retrieved from: ar2009en.pdf

Heller, D. (2017). The implications of digital currencies for monetary policy. In-depth analysis commissioned by the Directorate-General for Internal Policies. *Economic and Scientific Policy*, May 2017, 12. Retrieved from: http://www.europarl.europa.eu/regdata/etudes/idan/2017/602048/ipol_ida(2017)602 048_en.pdf

Ivashenko, A. I. (2016). Using cryptocurrency in the activities of Ukranian small and medium enterprises in order to improve their investment attractiveness. *Problems of Economy*, 3, 267–273.

Sauer, B., (2016). Virtual currencies, the money market, and monetary policy. International Advances in Economic Research, 22, 117–130.

9 Cyber security for sustainable construction

Introduction

Cyber security is the state of freedom from possible threats instigated by intentional, unwanted, ominous, or malicious acts. The security of any sector of construction is as important as putting the designs and plans into use in the first place. This encompasses all the activities involved in controlling, guarding, and managing measures put in place to safeguard the interest of the professionals and the project. It also relates to freedom from manipulations of data and information as well as cogent conditions in the contract during the tendering or procurement processes. The consequences of poor security are not to be taken lightly as they can be detrimental to the successful completion of the project. These consequences affect the profit and financial margin, business reputation, trust, scheduled programme, flow of work in respect of the duration, the building itself and, the most affected of all, the lives of the individuals involved. In relation to this, security issues have become a vital key in the Chartered Institute of Building (CIOB, 2018) Digital Special Interest Group (SIG). This was put in place to facilitate members to be security conscious by means of good and reasonable planning, and designing appropriate security measures for construction projects. The news of various breaches in security has placed the construction world at risk of private and confidential data falling into the wrong hands. Hence, the need has arisen to put various agencies and efforts in place in making sure these data are not exposed or compromised. Businesses nowadays are reliant on the Internet connections to transact; therefore, constructing firms need to be alert in preventing sensitive data from getting out without approval.

The role of cyber security in the construction industry

The need for cyber security

Cyber security can be defined as freedom from threats and harm as a result of interference, either deliberately or not. Cyber security can also be defined as the protection of computer systems from theft, damage, or corruption to

DOI: 10.1201/9781003179849-9

the software, hardware, and information. In this digital age, the need for cyber security grows increasingly as the need for computers grows as well. The construction industry is far from immune to cyber threats. The storage of critical data and information will always make it attractive to cyber criminals (Zurich, 2018). The construction industry holds vast amounts of information that is of interest to cyber criminals – from employee data to intellectual property – all of which can potentially be exploited for financial gain or other motives. For example, if a criminal gains access to the vital information, such as the design details for the construction of a revolutionary design or an audacious skyscraper, in order to announce his breaking and entering maliciously decides to change values of measurement. What fearful implications if this action is not noticed? It may affect the strength of the structure that would start developing faults that would cost a great deal in correcting, or worse, it may collapse. It could therefore lead to a high mortality rate and the contractor's organisation would be held liable for this disaster. Some criminals may not realise the damages they are about to cause as all they want to do is to earn brownie points for being able to break into certain levels of information, while other criminals may be terrorists who "broke in" intending to cause a collapse in the future. Cyber security helps to protect the database of information from such onslaughts (Schatz, Bashroush & Julie 2017; CIOB 2018).

One of the main features of the construction industry is that it is security minded (i.e. the privacy of contract), which includes not only the sharing of benefits but also the protection of confidential information. To be security minded means to understand and apply the appropriate security measures to avoid or prevent malicious activities (CIOB, 2018). As the world shifts to the usage of computer systems for the passage of information, for tendering, for communication and management of stakeholders, even for the mere signing of a contract, computer systems known for their storage capacities provide an avenue for the storing of confidential information. The CIOB, having realized the threats that a lack of security poses to the construction world, has set guidelines to help understand what constitutes good security, how to plan good security, and how to design mitigation measures for construction projects (CIOB, 2018). Cyber security helps to protect such information from the challenges of computer nowadays, one of which is cybercrime (e.g. theft and eavesdropping) (Tate, 2013). The consequence of poor security may be devastating to any project in terms of the profit, financial margins, the building itself, and at worse, risking the lives of occupants.

Construction firms involved in activities that facilitate a nation's strategic infrastructure (e.g. energy and transportation), and their increasing adoption of IT will make them prone to attacks. With the growing number of assets connected to the company's network, the division between the physical and the digital world grows thinner every day. Therefore, it is important to appreciate cyber risks and threats and carry out a risk assessment to know how to mitigate the threats. In addition, firms need the best cyber insurance policy so as to have the specialised experts required to help to manage this challenging area of risk.

Security can be assessed in the following four-fold methodology

- **People:** This includes the protection of humans within and outside of the organisation. It ranges from the protection of life to the protection of vital information concerning the personnel. For example, kidnapping of the project management on site will affect construction by causing a delay and extra cost.
- **Services:** These include the services delivered by the built asset, for example, avoiding tampering with a school website.
- **Access:** It is necessary to closely monitor the level of accessibility of varying calibres of people (strangers, acquaintances, customers, workers, and management).
- **Data:** The data to which varying personnel have access needs to be protected.

Some cybercrimes related to construction

- **Turner Construction (March 2015):** Employees of this firm were scammed by a fraudulent email account that also deceived the employees. Their firewall was breached and information was stolen. The employees in the company in 2015 at the time of the breach time were implicated. The company was advised by an outside vendor that prepared W-2 and 1095 tax forms for the company's employees about suspicious activity on that vendor's systems.
- John Smith was a manager at Bid Old Construction. He received an email supposedly from his boss, the company's CFO, Jane Doe. Jane sent instructions to John to send $10M USD to an account in Geneva to complete a deal that the Board had authorised. John obeyed the instruction as he found nothing suspicious about the instruction, only for his boss to find out that they had been scammed.

Basic security protocols

The risk of cybercrimes can be mitigated using the following approaches.

Be security conscious

This applies both professionally and personally as an individual or an expert. Managers are expected to be enlightened about cyber threats. They should be able to identify possible threats, assess the chances of that happening as well as its impact, and then suggest measures to prevent them.

Compliance with professional ethics, legislature and codes of conduct

Managers are expected to be aware of the legislative laws of the country in terms of security and adopt them. They need to employ personnel who are skilled in this area.

Communication

It is essential to be security conscious in communication, to create an awareness of the need for security, and to make available a report system for potential threats or suspicious activities.

Awareness

Knowledge in terms of security awareness needs to be shared with the employees. In addition, security alertness implies sensitivity to the environment and an awareness of what is happening around them.

Insurance

Most insurance companies embark on using the online platform as a way of managing risk. They also use it to assist the insured to attend to matters of security or personally inflicted incidents. This gives them the benefit to access free loss prevention tools and training. This is just a transfer of risk if an assessment reveals that a firm is prone to these attacks.

Hiring a security expert

A computer expert control units are present in the system in order to have a protected and functioning system. This expert is saddled with all-round responsibilities of ensuring functionality of a designated system.

Cyber security measures for sustainable construction

Adopt a security-minded approach to your professional and personal life

Being informed about the way various actions affect security when the influence of social media is also taken into consideration.

1 Identifying the vulnerabilities and other deficiencies at each stage of the construction implementation, and suggesting preventive or corrective solutions to mitigate risks that could have been avoided;

2   Taking into consideration the failure to deal with the identified vulnerability will have a negative impact on what is subjected to it;
3   Comprehending the need for bringing in several personnel, and practising all forms of security that could be employed;
4   Demonstrating capability in moderating the contents that are available through the media platform without exposing too much on the Internet; and
5   Identifying and overseeing the execution of fiscal policies that are currently relevant or on a permanent base to help in any phase of the construction work.

Legislation and codes and further improvements

- Being made to follow the laws that govern a particular region in which the operation is being carried out strictly without diverging from any of them under any circumstances and
- Not ignoring the codes of conduct that relate to the practices of the locality in which the construction is being carried out.

Ensure good security-minded communications

- Be able to identify the security policies and processes relevant to staff and other members of the supply chain, and communicate them clearly and effectively;
- Be able to express clearly the balance of security risks and opportunities;
- Adopt an 'open reporting' approach to security risks, incidents and near-misses, coupled with a spirit of questioning and learning from others; and
- Be selective of the material used when publishing information at conferences, workshops, and seminars or when writing in professional or trade publications to avoid releasing sensitive data and information.

Improving lasting systems for security governance

- Demonstrate understanding of one's personal role in contributing to the security of the built asset at the time;
- Making contributions to the development, implementation, and review of security policies and processes that are concerned at that particular moment;
- Ensure security-related roles and responsibilities are clearly designated and understood by staff and members of the supply chain;
- Improving one's knowledge and understanding of security risks and establishing curbing measures that can be applied during construction;

- Contributing to the development and implementation of appropriate mechanisms for reporting and feedback on security incidents and issues; and
- Contributing to the scrutiny and auditing of the security culture and implementing security policies and processes.

Causes of incidents in cyber security

- Compromised data via a phishing scheme, lost documents (tangible and intangible), or laptops;
- A hacker aiming to make money or cause harm to your organization; and
- The system itself – breach in the hardware or security system can be exploited by a third party.

Contribute to public and professional awareness of security

- Be able to engage in debates on security risks and benefits, especially in relation to new technologies and innovative developments;
- Be able to recognise the social, political, and economic implications of security risks and acknowledge these through appropriate channels;
- Be honest and clear about uncertainties, and be prepared to challenge misrepresentations and misconceptions; and
- Contribute to public and professional awareness of security by appropriate sharing and promoting knowledge of effective solutions.

What information could be exposed in a breach of cyber security?

- Log-in credentials in order to access customer systems;
- Financial banking information of your customers and employees;
- Intellectual property of your customers, such as drawings or specifications; and
- Other personally identifiable information such as social security numbers, names, and addresses.

What you should be doing about it

Culture: Engaged employees

- Beware of phishing schemes – review the sender's address before clicking on links (when in doubt, don't click);
- Encrypt laptops and implement a clean desk policy to ensure safety measures when not around; and
- Be cautious of allowing entrance to your building.

Cyber security mitigation platform

- Sticking by the industry standard patching cadence for antivirus, operation systems, and software;
- Proactive IT department – educating employees on how to identify phishing schemes and other attempts to gain access to your systems;
- Data assessment and profiling to ensure you only keep the information you need; and
- Robust data encryption approach – using of strong password requirements with frequent changes.

Cyber coverage for construction industry

The solidarity of the construction industry becoming connected through Internet-connected solutions (collectively storing and sharing of information) and remotely accessible systems give platforms for hackers to establish a cyber attack. Any company that utilizes email or Internet-related packages for business is at risk of a cyber event as explained earlier. Attackers are becoming more sophisticated and are aiming for companies like those in construction that many assume to be safe.

- Construction firms keep a variety of information that might be of interest to hackers in terms of:

  - architectural drawings and specifications,
  - proprietary assets,
  - intellectual property,
  - building schematics or blueprints,
  - and more.

- Construction firms have valuable details about their clients (e.g. corporate banking and financial account information) all of which are prime targets for attack.
- Construction firms are frequently targeted with spear-phishing campaigns looking to gain access to data that are registered under the employees' data such as full names and addresses, social security numbers, and bank account details used for payroll.

Construction industry instances of cybercrime

Case study one

Company size: $200M in corporate programmes per year and $100–150M in projects per year.

In a large civil construction company in a Gulf Coast state, hackers took total control of all systems (Email both internal and external, payroll, phones, and accounting) and demanded $250k in bitcoin. After months of negotiating, they ended up paying $150k in bitcoin to get their systems back. The client had been emailing on personal Gmail accounts. Multiple NOCs went out as the client did not have access to accounting and could not pay the premiums. The total downtime was between two-and-a-half to three months. The client rejected cyber before International Risk Management Institute (IRMI), declined the cyber meeting with the markets at IRMI, and rejected coverage at renewal, and roughly 2–3 weeks later, the company was hacked.

Case study two

John Smith, controller for Big Old Construction, received an email from the company's CFO, Jane Doe. Jane sent instructions to John to send $10M USD to an account in Geneva to complete a deal that the Board had authorized. At first glance, John saw nothing out of the ordinary with the word pattern in the email and the email address looked correct. John sent the funds in accordance with Jane's wishes. Upon Jane's return to the office, John realized the email was a hoax.

Sustainable construction

Sustainability, sustainable development, green building, eco-friendly... These are commonly used words in the construction sector today. Sustainability in construction is one of the 21st century developments that have turned out to be one of the major drivers of the modern day construction. Attempts have been made to define what sustainable construction is, its scope, where it starts, and where it ends. While the definition still needs to be improved upon or debated, we can, however, define what sustainable construction is about. Sustainable construction is about the production of eco-friendly construction. The idea of sustainability is centred on saving the environment from the perils of development even though it varies from place to place. Sustainability in building looks to reduce the adverse health and environmental impact of construction or construction-related activities. We can deduce that sustainable construction is all about the protection of life and the environment with a view to minimising waste (Bourdeau, Huovila & Lanting, 2010; UNDP, 2015). Sustainable construction, according to CIB (1994) and Kilbert (1994), is the creation of responsible management of healthy building practices on the basis of ecological principles and resources management efficiency.

Though, there has been the need for sustainability since the beginning of time, since the definition has made us realise that it focuses on saving the earth, which is the inhabitant of every life form, from the dangers of change, development, and by extension, mankind. It was not until 1970 that the people realized the urgent need for energy conservation. This was triggered by the oil crises (BIS, 2010). Sustainability entered into construction in the 1960s with the aim of seeking harmony with nature.

Benefits of sustainable construction

Sustainability is a dimensional concept. The concept of sustainability could be explained in two parts. The first is concerned with the planet itself with little or no regard for the social issues or the happenings on earth, the war, and human activities. It is all about saving the earth and keeping earth pure, amongst other current movements in the name of sustainability. The second relates to development and tries to look at sustainability from the perspective of things to do to ensure the continuing economic growth of a society that ignores the ecology. Some have tried merging the two perspectives, thereby causing a more complex problem (Khan, 2002).

The economic perspective is concerned about the extent to which we can limit growth so as to preserve the economy for future generations. The ecological perspective, on the other hand, concerned about keeping the earth clean and free from toxins that could threaten our existence. The ecologists hold on to the belief that any human being is entitled to a good quality of life and not just a biological survival. Therefore, concerns are raised regarding energy and energy consumption, both of which they are critically monitoring. Statistics have shown that with the increased level of construction activities globally, there will be increased urbanisation that in turn will lead to an increase in waste and pollution and as a result, the depletion of natural resources and wildlife habitat by 2032.

The effects of sustainable construction

1 The cost of construction and setting up of a sustainable environment might be relatively higher; however, construction costs will ultimately be reduced with the invention of sustainable technologies (Pantouvakis & Manoliadis, 2006).
2 Sustainable construction will facilitate better designs in the construction phase (Shealy, 2016).
3 The environment will be protected as contractors would employ the use of management plans to combat the effect of projects on the environment (Manoliadis & Pantouvakis, 2006).
4 The promotion of sustainability will increase the awareness that sustainability is important for life (Pantouvakis & Manoliadis, 2006; Shealy, 2016).
5 The promotion of sustainability will also enable the market for sustainable products.
6 Improving healthy living and reducing disease will result from sustainable construction (Shealy, 2016).

7   A further benefit is the reduction and minimisation of energy consumption.
8   Building sustainably will improve the relationship between shareholders and the local communities.
9   It will help to promote cost effective and socially acceptable project plans.
10 Sustainable construction builds a better future for future generations, or at least protects the present environmental conditions.
11 Building sustainably facilitates and enhances the use of renewable resources.

Information technology and sustainable construction

Promoting sustainability is the right way to ensuring of the future of humanity and quality of life on earth (Khan, 2002). Irrespective of the technological development, supercomputers, humanoids, and artificial intelligence (AI), the issue of sustainability will remain a focal point, even in the future of technology. This is because we can see that sustainability is at the heart of our society. Information technology (IT) is one of the strategic drivers of sustainability. Investment has been made into technological advancements that have brought about efficiency in construction technologies and methods for reducing construction waste and saving time (Jones, 2018). IT has modelled reality in such a way that information can be transferred efficiently to management in a useable and readable form (Khan, 2002). There is still the challenge of standardisation of the language or formats interchange that will help enhance communication. IT has changed our reality; standardisation issues may alter such realities that will in turn lead to unsustainable decisions (Khan, 2002).

The role of information technology (IT) in achieving sustainable construction

Technological change is penetrating into all aspects of construction, from the use of digital equipment for planning, designing, costing or even construction to the use of laptops and smart phones (for communication). The Internet is fast becoming one of the most essential and sought after infrastructures of this age (World Economic Forum, 2014). It has taken about two decades for the commercial Internet to rise from just a fascinating innovation to an indispensable tool, from excitement (fun) to an essential. Statistics have shown that about 2.5 billion people are on the Internet daily (World Economic Forum, 2014). It is then safe to say that the Internet has fast become a fact of life.

Technological advancement has always driven construction, even though the level of the adoption of IT in construction is way below the level of investment into the development of more sophisticated machineries and IT generally for the purpose of construction (Jones, 2018). This has made it far from being immune to cyber threats. The construction industry is an industry that operates on the tangible, making it difficult for firms to fully recognise the potential for cyber threats and attacks (Zurich, 2018). New technologies in construction are developed at such a speed that what seems like a 2050 technology is brought to life in front of our very eyes.

Conclusion

In everything that is being done and estimated for, the security of the content is always a priority to whatever concerned party. Security cut across every facet of human survival as it is expressed in social, physical, and so many others. The amount of data and information involved in construction has to be kept safe to keep sanity and order in the progress of the work estimated for from the onset of the project. Cyber security towards sustaining the parts involved in construction brings an umbrella that shades the confidential documents that could affect the course of the project. This could place the client, contractor, and other persons involved in the project in a difficult and unplanned for situation if the information eventually gets into the wrong hands through incompetence or hacking. Cyber security helps to protect both the interests of the project and the construction professionals involved.

References

Bourdeau, L., Huovila, P. & Lanting, R. (2010). Sustainable development and the future of construction. *Proceedings of the CIB W82 Project Conference, Amsterdam*, 14(2), 1–8.

Chartered Institute of Building (CIOB). (2018). *The role of security in the construction industry*. Designing Buildings. London UK.

Jones, K. (2018). *How technology is reshaping the construction industry*. Retrieved from ConstructConnect Blog: constructconnect.com

Khan, H. A. (2002). *Sustainable development and information technology*. Retrieved from: http://www.sciencevision.org.pk/Backissues/Vol7/Vol7No1-2/Vol7No1&2_2

Kilbert, C. J. (1994). *Establishing principles and a model for sustainable construction*. Retrieved from: https://www.irbnet.de/daten/iconda/CIB_DC24773.pdf

Manoliadis, O. G. & Pantouvakis, J.-P. (2006). A practical approach to resource-constrained project scheduling. *Operational Research*, 24(2), 299–309.

Pantouvakis, J.-P. & Manoliadis, O. G. (2006). The business case for green building. *World Green Building Council*, 6(3), 299–309.

Schatz, D., Bashroush, R. & Julie, W. (2017). Towards a more representative definition of cybersecurity. *Journal of Digital Forensics, Security and Law*, 12(2), 53–74.

Tate, N. (2013). Reliance spells ends the road for ICT amateurs. *The Australian. Computer security and mobile security challenges* (pdf). https://researchrepository.murdoch.edu.au.

Shealy, T. P. (2016). Do sustainable buildings inspire more sustainable buildings? *Procedia Engineering*, 145(2016), 412–419.

United Kingdom Department for Business, Innovation and Skills (CIB). (1994). Sustainable construction. In: *Proceedings of the First International Conference of CIB TG 16, Florida, USA*, 12(55).

United Kingdom. Department for Business, Innovation and Skills (BIS). (2010). Estimating the amount of CO_2 emission that the construction industry can influence. *BIS*. Retrieved from: https://www.gov.uk/government/organisations/department-for-business-innovation-skills

United Nations Development Programme (UNDP). (2015). Sustainable development goals report. *United Nations Sustainable Development Summit*, 25–27 September, New York.

World Economic Forum (WEF). (2014). Delivering digital infrastructure, advancing the internet economy. *World Economic Forum, Geneva*, June 2014.

Zurich. (2018). *Cybersecurity and the construction industry: Staying ahead of emerging threat*. Retrieved from: https://www.zurichna.com/knowledge/articles/2019/08/cybersecurity-and-the-construction-industry

10 Drone and sustainable construction

Introduction

The world has experienced a great deal of advancement in information and communication technology (ICT) in recent times. The 21st century earmarked the willingness of humans to learn more about the efficiency and effectiveness of machines and technology. Some of the most recent innovations in technology now being applied to many sectors and industries are robotics, mechatronics, machine learning, artificial intelligence, nanotechnology, the Internet of Things (IoT), cloud computing, drone technology, and augmented reality. The common ground for all these concepts is the human input of programming and coding and the automation of electronic devices using a remote computer. The industrial sector has been the forerunner in adopting and applying these innovations; however, other sectors have also adopted these principles, especially in national security. Recently, the construction industry has been applying these concepts in a bid to increase the efficiency and better management of the construction process.

Definition of terms

Drone

A drone is otherwise referred to as a unmanned aerial vehicle (UAV). As the name implies, it refers to a remotely operated aircraft that has no human on board as the pilot, although it is still being controlled by a human. UAVs function mainly with the help of a system of communication with a ground-based controller. In other words, drones are flying robots that can be remotely or autonomously harnessed with the aid of embedded controlled flight plans that are controlled by the software, working with corresponding on-board sensors, and a global positioning system (GPS).

Sustainability

In its simplest form, sustainability is the ability to exist perpetually; it is also the enablement to be maintained at a certain rate or level without diminishing.

DOI: 10.1201/9781003179849-10

The Oxford dictionary defines sustainability as the avoidance of the depletion of natural resources in order to maintain an ecological balance. When something continues to exist over a long period or many cycles, without needing external resource input, it is deemed to be sustainable.

Sustainable construction

Sustainability is definitely amongst the most discussed, yet least comprehended terms of the 21st century. Its essence is regularly obfuscated by varying understandings exacerbated by an inclination to treat the topic externally, be it through 'eco', 'green', or 'brilliant' talk. However, for those who take the issue seriously, the word 'sustainability' pertains to the earth and its long-term endurance as an issue of concern for all mankind. The United States of America's Environmental Protection Agency (EPA) defines sustainable construction as the act of making structures and utilizing forms that are earthly mindful and asset proficient all through a structure's lifecycle from setting to design plans, development, activity, support, remodelling, and the eventual deconstruction. In Holland, sustainable construction is explained as a method for infrastructure development that aims at reducing negative well-being and harmful ecological effects brought about by development, by structures, or by the built environment. The major recurring concepts as identified in sustainable construction are protecting the natural environment (environmental), selecting non-toxic materials, reducing, and re-using resources (technical); minimising waste (social), and using of lifecycle cost analysis (economic and financial). Considering the global extent of urbanisation today and the pace at which the planet is being urbanized, it is much basic that whatever that is manufactured must perform reasonably on all registers – environmentally, financially, and socially.

Despite the fact that progress is inevitable, the discussion proceeds with respect to whether it ought to continue: radical or steady change – and that always possess issue. However, whatever the appropriate response (either fundamental change or tweaking business as usual), we do not have the advantage of time to choose the appropriate mode. Sustainable development, in accordance with the stipulations for advancement delineated in the Brundtland Commission's report 'Our Common Future' from 1987, set the standard that is acceptable today in terms of accommodation, expected and unexpected working conditions, along with proper foundation without deviating from the concept of reuse in the future to address issues for safe houses, spaces for work, and arrangement.

Drone technology

Siebert and Teizer (2014) identified different names for drones as mentioned earlier, names such as remotely piloted vehicles (RPVs), an unmanned aerial system (UAS), and unmanned aerial vehicles (UAVs). Drones have enjoyed continued application in numerous industries over the years and are now being employed in the construction industry (Dupont et al., 2017). Drones were initially applied to military deployments. They possess a significant situation in the military munitions stockpile (Holton, Lawson & Love, 2015). However, costly military automatons are frequently not easily available for some clients in common application, for example, for amusement and transportation. These days, non-military automatons have enabled 3D top notch mapping information to be significantly more open. Drones can be ordered in many forms such as multi-rotor, fixed-wing mixture, fixed-wing, and single-rotor (Australian UAV, 2017). Among these, multi-rotor automatons such as quadcopters are identified with operations that are favourably compared with other UAV frameworks. The examples are power, high movement, and low purchase and maintenance costs. Multi-rotor automatons have multiple rotors and use fixed-pitch cutting edges. Controlling the movement of the vehicle is done by shifting the total speed of every rotor to alter the force in respect to the push, torque delivered by each vehicle. Along these lines, many fields are demonstrating a wide range of enthusiasm for using multi-rotor rambles for different non-military purposes (Irizarry & Costa, 2016). For instance, in ranger service and horticulture, to accomplish site-specific in weeding the board, the ultra-high spatial and high phantom goals symbolism gave by multi-rotor automatons were utilized for weed mapping at early phonological phases of harvest and weed plants. In crisis and fiasco on the board, multi-rotor automatons were applied to scan for and salvage individuals caught by garbage or harmed during catastrophes (Erdelj, Krol & Natalizio, 2017). In traffic monitoring, an automaton-based vehicle identification framework was created by Wang, Chen and Yin (2016) to gather traffic data, track vehicles, and screen driver behaviour. Significant tests exhibited accurate precision in vehicle location and identification. What is more, past investigations likewise widely investigated and created UAV photogrammetry for 3D mapping and demonstrating (Nex & Remondino, 2014; Martin et al., 2016; Trujillo et al., 2016).

Application of drone technology in construction

Multi-rotor drones as a creative innovation in the modern construction environment can possibly encourage development projects from inception to observing of safe works, quality time, cost saving, and prevention of hazards as well as prompting quality work (Herrmann, 2016). Be that as it may, multi-rotor automatons have not been generally utilized in the development business as it has been a moderate adopter of upcoming innovations (Holt, Benham & Bigelow, 2015). Thus, generally little research has focused on the potential uses of multi-rotor rambles in development designing and the executives when compared with some other fields. It is advantageous to bring issues to light of the utilization of multi-rotor rambles by examining the assistance that they can bring to the present and future development in construction industry.

Site survey

Surveying and geographical mappings are major parts of all infrastructural development ventures and a key system toward the commencement of the development procedure. Conventional land surveying strategies require massive apparatuses, for example, tripods, total stations, and GSP hardware. Multi-rotor rambles outfitted with picture-handling programming, cameras, and autopilots can be applied to land surveying and mapping in development ventures. Planned with fitted sensors and camera advancements, UAVs give an accurate platform for obtaining land information and any other additional data needed for present or for future reference. The examination of Siebert and Teizer (2014) brought about two sorts of estimations; a manual, ground-based, ongoing kinematic GPS study, and an automaton-based photogrammetric overview close to the city of Magdeburg, Germany. In addition, Fleming et al. (2016) chose the Transbay Transit Centre building site with measurements of $457 \times 56 \times 20$ m in downtown San Francisco as a contextual investigation.

Inspection

It has been approved in the oil business and conglomerate that the application of UAVs can improve the pipeline investigation process and well-being by decreasing labourers' exposure to dangerous atmospheric condition without having proper knowledge of possible hazards. In terms of the construction business and industry situation, a UAS could examine an outside water spill on an upper floor of a tall structure where a window casing is colligated to be the source with the break. The structure is placed at sides by a bustling expressway, granting access by different methods. UAS could take various photographs from different angles and enlarge settings in mere minutes. The entire activity could be done without street closures, or the establishment of a suspended framework, at an incredibly lower cost and with next to no danger (Pritchard, 2015).

Safety management

Maintaining security is a prominent issue in research and related practices, and every now and then, labourers or unskilled workers meet with deadly mishaps in the construction industry (Chen, Song & Lin, 2016; Enshassi, Ayyash & Choudhry, 2016; Park, Kim & Cho, 2017). Irizarry, Gheisari and Walker (2012) at first explored the potential benefits of automaton-related advancements for well-being administrators in the construction industry as factors were reviewed to give an exclusive understanding of the effects of safety. As multi-rotor automatons can gather and convey ongoing recordings of the present circumstances at construction sites. A test was drawn up by the analysts to reproduce a construction site. The examination tests whether labourers were wearing their hard hats and other safety measures in order to establish adherence to safety protocols by using iPads and iPhones attached to drones. The test was set in a view to capture workers in hard hats and crosscheck visions supplied. The outcomes revealed the practicability of multi-rotor rambles as it demonstrates that both apparent view and iPad perception conditions could result in an acceptable exactness in hard hat recognition with no problem whatsoever.

Quality management

Rumane (2016) noticed the more need to consider quality management in construction. Quality management has gained recognition over the years due to its growing additions to the construction industry. Specifically, construction faults are the essential drivers of low levels of efficiency, delays, extra expenses, and the requirement for additional materials and labourers for imperfection rectification (Aichouni et al., 2014; Lee et al., 2016). Therefore, adequately distinguishing imperfections right at the outset in the development procedure is a basic way of maintaining standard control. Building information modelling (BIM) can likewise give accurate information about a project to be executed through its integrated systems of operations (Hardin & McCool, 2015). Automaton innovations are designed to assist in working output and improving the efficiencies in undertaking checking and quality control for BIM-related development ventures (Tezel & Aziz, 2017). Wang et al. (2015) identified ways of controlling operations in construction by distinguishing the practices employed locally and that of the international by using BIM and LIDAR. Their work exhibited an incorporated arrangement of BIM and LIDAR towards accomplishing development quality control. The BIM–LIDAR approach depends on a LIDAR-framework, a BIM-based ongoing checking framework, a quality control framework, a point cloud change framework, and an information handling framework. To benefit quality innovators in quality evaluation and quality control, a change module for automaton flight was utilized to change predefined flight way parameters into an automaton flight way control framework. Therefore, the multi-rotor drone will be accomplished a flight way equivalent to a predefined flight way in the virtual condition. BIM was additionally demonstrated to be a fruitful representation stage and benchmarking model in the quality administration test that added more relevance to it. The contextual analysis exhibited an improvement to tedious quality investigations.

Time management

Notwithstanding overall workers' well-being and quality reviews, construction duration can be improved by automaton BIM advancements in development ventures (Vacanas et al., 2016). 3D BIM models can be improved and updated whenever connected with timetable (4D), costs (5D), and undertaking lifecycle data

(6D) (Park & Cai, 2017). BIM models are now common in various applications of monitoring construction progress by introducing multi-dimensional information (Chen & Luo, 2014). Multi-rotor automatons plan to efficiently gather records and as-manufactured data, even in indoor building locales (Hamledari, McCabe & Davari, 2017; McCabe et al., 2017). BIM can be then refreshed to evaluate the data whether it will bring about any postponements due to possible variance in outcome analysed. Irizarry and Costa (2016) mentioned four case studies to recognising uses of multi-rotor rambles during the development procedure. The first task explained the demolition and renovation work for an office in Atlanta City, Georgia, USA. The other two tasks were the development of a school building research and a secondary school in Georgia. The last task was identified with the development of eight loft structures in the city of Salvador, Bahia in Brazil. A total of 200 visual resources including 98 photographs and 102 recordings were collected inside a 7-month time frame through automaton flights at the places of work. In contrast to manual methodology, the incorporated innovation of BIM, UAS, and continuous cloud information enables quick ongoing task control, observing, and review by looking at pre-arranged data and as-assembled conditions of the development ventures (Ham et al., 2016). Lin, Han and Golparvar-Fard (2015) proposed a model-driven methodology for obtaining and investigating progress pictures. Automaton-based advancement observing and transient data in 4D BIM for independent information obtaining were discussed.

Demolition management

At present, drones could be deployed for waste administration (Ge et al., 2017) and to make recordings to determine the progress of the demolition (Parramatta Advertiser, 2017; Taylor, 2017). For instance, O'Neill (2016) made an amazing film of a controlled emergency clinic demolition of the 11-story Millard Fillmore Gates Hospital in New York, from the perspective of a flying automaton: the structure was razed to the ground inside 30 seconds. The captured video showed information gathered by the deployed drones and thus was utilized for development and post-development stages. The blend of automaton and point cloud advances is by and large broadly misused by specialists in the construction industry. For example, in land surveying, observing progress and basic assessment, the information collected is frequently sent out into BIM models to accomplish the quality, progress, and waste administration.

Site management

In an attempt to better the normal routines of project site management, increased perspectives on building locales were accommodated by development designs regarding high perspectives and a mix of real and virtual scenes (Wen & Kang, 2014). 3D portrayals utilizing augmented reality (AR) innovations were created from pictures that automatons captured at specific heights and areas. The proposed strategy of joining AR and automaton-related advancements can help experts in imagining both real field and virtual development situations. It would empower supervisors to design parts of the building site in order to recognize potential issues.

Environmental impact

This perspective looks at sustainable construction as including the plan (design) and the effective monitoring and management (in value) of constructed structures, regardless of the size of structures, foundation, or urban agglomerations; the availability of materials over all scales and all through their entire use-cycles; and the utilization of sustainable power source assets in structure, activity, and support to reduce global harmful gas emissions.

Economic impact

This sees sustainable construction as making progress from a state of economy to a better and advanced one that encompasses sustainable power source usage, material and waste reusing, water gathering and safeguarding, transferable data, and the flexibility of structures to changes being used; creative financing models based on an economy of yielding more with less with the aim for further investment.

Social impact

Economic development includes adherence to the most noteworthy moral models in business and industry rehearses all through all task stages; the advancement of socially feasible living and workplaces, including word-related well-being and security benchmarks for work powers and clients; and the democratization of all procedures relating to the generation and utilization of the manufactured condition as a region.

Technical impact

Very recently, an additional dimension has been added to the initial three components of sustainability listed earlier. Technical sustainability has to do with technological innovations in construction. Simply put it is the use of locally developed technologies that continue to function for a lengthy period and are maintained (repaired and replaced) by the local people.

Drone technology for sustainable construction

The construction industry keeps on investigating the utilisations of drones since construction organisations and design specialists have identified the importance of automatons utilisation in executing construction work. As depicted in the past segments, it becomes imperative to draw a line as to how drone technology promotes sustainable construction.

Environmental perspective

The use of drones in construction is as harmless as possible. There is no interference with the environment as it doesn't emit carbon monoxide that is toxic. Its operations are within a programmed units functioning together in components without causing havoc whatsoever.

Social perspective

The view from social perspective is that the fundamental commitment of multi-rotor drones on project sites is directed towards determining work security and safety challenges or issues. For instance, land surveyors more often than not work in a risky domain because of exceptionally slanted surfaces or being near overwhelming gear. Their work is always outside, paying little heed to climate conditions (El Meouche et al., 2016). Having an automaton mapping arrangement takes into account, self-sufficient flight to dispose off a few dangers related with land surveying, for example, substantial hardware and damage from risks. Automaton-based advancements can likewise resolve the issues of difficult and hazardous auxiliary assessments, for example, those of leaky rooftops, outside veneers and dividers, towers and extensions, damage due to fires and blasts, vehicle accidents, and calamitous occasions (Mat Yasin, Zaidi & Mohd Nawi, 2016). Anticipating how structures will collapse can guarantee a safe demolition condition by means of recreation, in which 3D building models can be created through structure warm reviews from automatons. In general, damage and casualties can be significantly diminished or avoided through the viable utilization of multi-rotor rambles.

Economic perspective

As mentioned earlier, drone-based reviewing methodologies are generally savvy. Multi-rotor automatons can accomplish the fast gathering and programmed investigation of territory information. Automatons can likewise be utilized to robotize other straightforward errands and significantly lessen task costs. Rather than utilizing HR, and costly reviewing instruments, ramble-based advances are equipped for creating complex information with less cost and more exact precision (Siebert & Teizer, 2014). Basic reviews regularly require run-of-the-mill assessment units, for example, truck cranes, hoisting stages, and under-connect units (Morgenthal & Hallermann, 2014). Automaton-based assessment methodologies can likewise avoid the high strategic and staff costs that enormous trucks, uncommon hoisting stages, and frameworks require. Moreover, ramble mapping is likewise unsurpassable in terms of speed compared with customary methodologies of land surveying. Conventional land may require extended periods of time and the conveying of substantial gear, starting with one area then onto the next. Be that as it may, ramble mapping may require just minutes to finish a site overview with higher precision, rather than days or weeks. Automaton-based technology can help to avoid postponements in development ventures. A postponement to fulfilment and conveyance can bring about additional expenses and decrease profitability because of noteworthy misfortunes and costs.

Conclusively, building analysts have additionally been investigating new route frameworks for advancing automaton advances, which implies that future automatons will have the option to withdraw from route reliance on GPS satellites (Stark, 2017). Automatons will be able to explore self-sufficiently inside the structures and framework under development, for example, structures in gulches, underground, and in different spots where GPS signals are inaccessible or untrustworthy. In this manner, the new innovation will accomplish quality investigations and time would be conserved for other aspect of the construction. Moreover, fatigue is expected to be exhibited by both the workers and supervisors when large construction is to be executed (Jarkas, Balushi & Raveendranath, 2015). It very well may be anticipated that automatons fitted with exhaustion identification frameworks will have the option to screen numerous vehicle and gear administrators' facial expressions to decide whether they are in danger of nodding off on a building site, and this will reduce the workload on both the supervisor and the workers involved in the project.

Conclusion

Drone technologies over the years but it is obvious that it has come to stay. Not only does it come with more fanciful designs in various shapes, but it also comes with upgraded versions that could cater for a detected inadequacy in the previous versions. It is quite economical and the efficacy associated with it improves the image of the construction firm that uses it to a higher level than the rest. The application of drone technology has brought more advantages than the feared consequence of creating unemployment by getting rid of human labour (or with little participation). It has eased the tension in some aspects of construction in terms of speed and accuracy. Sections of works are carried out effectively and correspondingly without the fear of cost overrun and wastage of materials. Drone usage is definitely heading the construction industry towards an acceptable standard through its various functions and advantages. It has come to stay, and the challenges that many are still concerned about will not be able to stop it.

References

Aichouni, M., Ait Messaoudene, N., Al-Ghonamy, A. & Touahmia, M. (2014). An empirical study of quality management systems in the Saudi construction industry. *International Journal on Construction Management*, 14(3), 181–190.

Australian UAV. (2017). Types of drones: Multi-rotor vs fixedwing vs single-rotor vs hybrid VTOL. *Australian UAV*. Retrieved from http://bit.ly/2xPV1be

Chen, J., Song, X. & Lin, Z. (2016). Revealing the 'Invisible gorilla' in construction: Estimating construction safety through mental workload assessment. *Automation Construction*, 63, 173–183.

Chen, L. & Luo, H. (2014). A BIM-based construction quality management model and its applications. *Automation Construction*, 46, 64–73.

Dupont, Q. F. M., Chua, D. K. H., Tashrif, A. & Abbott, E. L. S. (2017). Potential applications of UAV along the construction's value chain. *Procurement Engineering*, 182, 165–173.

El Meouche, R., Hijazi, I., Poncet, P., Abu Nemeh, M. & Rezoug, M. (2016). UAV photogrammetry implementation to enhance land surveying, comparisons and possibilities. In *International Archives of the Photogrammetry, Remote Sensing and Spatial Information Sciences*, Copernicus GmbH, Gottingen, XLII-2/W2, 107–114.

Enshassi, A., Ayyash, A. & Choudhry, R. M. (2016). BIM for construction safety improvement in Gaza strip: Awareness, applications and barriers. *International Journal of Construction Management*, 16(3), 249–265.

Erdelj, M., Krol, M. & Natalizio, E. (2017). Wireless sensor networks and multi-UAV systems for natural disaster management. *Computer Networks*, 124(4), 72–86.

Fleming, K. L., Hashash, Y. M. A., McLandrich, S., O'Riordan, N. & Riemer, M. (2016). Novel technologies for deep-excavation digital construction records. *Practical Period Structure Des Construction*, 21(4), 05016002-1-10.

Ge, X. J., Livesey, P., Wang, J., Huang, S., He, X. & Zhang, C. (2017). Deconstruction waste management through 3D reconstruction and BIM: A case study. *Visual Engineering*, 5(1), 13.

Ham, Y., Han, K. K., Lin, J. J. & Golparvar-Fard, M. (2016). Visual monitoring of civil infrastructure systems via camera-equipped unmanned aerial vehicles (UAVs). A review of related works. *Visual Engineering*, 4(1), 1–8.

Hamledari, H., McCabe, B. & Davari, S. (2017). Automated computer vision-based detection of components of under-construction indoor partitions. *Automation Construction*, 74(Supplement C), 78–94.

Hardin, B. & McCool, D. (2015). *BIM and construction management: Proven tools, methods, and workflows*. Indianapolis, IN: Wiley.

Holt, E. A., Benham, J. M. & Bigelow, B. F. (2015). *Emerging technology in the construction industry: perceptions from construction industry professionals*. 2015 ASEE Annual Conference & Exposition, Seattle, WA.

Holton, A. E., Lawson, S. & Love, C. (2015). Unmanned aerial vehicles. *Journal Practical*, 9(5), 634–650.

Hubbard, B., Wang, H., Leasure, M., Ropp, T., Lofton, T., Hubbard, S. & Lin, S. (2015). Feasibility study of UAV use for RFID material tracking on construction sites. Proceedings of the 51st ASC Annual International Conference, College Station (TX), The Associated Schools. *International Journal on Remote Sensing*, 38(810), 2161–2176.

Irizarry, J. & Costa, D. B. (2016). Exploratory study of potential applications of unmanned aerial systems for construction management tasks. *Journal of Management Engineering*, 32(3), 05016001-1-10.

Irizarry, J., Gheisari, M. & Walker, B. N. (2012). Usability assessment of drone technology as safety inspection tools. *Journal of Information in Technological Construction*, 17, 194–212.

Jarkas, A. M., Balushi, R. A. & Raveendranath, P. K. (2015). Determinants of construction labour productivity in Oman. *International Journal of Construction Management*, 15(4), 332–344.

Lee, D.-Y., Chi, H.-L., Wang, J., Wang, X. & Park, C.-S. (2016). A linked data system framework for sharing construction defect information using ontologies and BIM environments. *Automation Construction*, 68, 102–113.

Lin, J. J., Han, K. K. & Golparvar-Fard, M. (2015). A framework for model-driven acquisition and analytics of visual data using UAVs for automated construction progress monitoring. In *Proceedings of the International Workshop on Computing in Civil Engineering*, 21–23 June, Austin, TX, 156–164.

Martin, P. G., Kwong, S., Smith, N. T., Yamashiki, Y., Payton, O. D., Russell-Pavier, F. S., Fardoulis, J. S., Richards, D. A. & Scott, T. B. (2016). 3D unmanned aerial vehicle radiation mapping for assessing contaminant distribution and mobility. *International Journal Application Earth Observation Geoinformation*, 52, 12–19.

Mesas-Carrascosa, J. F., Rumbao, C. I., Sanchez, J. T. & Ferrer, A. G. (2017). Accurate ortho-mosaicked six-band multispectral UAV images as affected by mission planning for precision agriculture proposes. *International Journal of Remote Sensing*, 38(8–10), 2161–2176. Doi: 10.1080/01431161.2016.1249311

Mat Yasin, M. F., Zaidi, M. A. & Mohd Nawi, M. N. (2016). A review of small unmanned aircraft system (UAS) advantages as a tool in condition survey works. *MATEC Web of Conferences*, 66(2016), 00038.

McCabe, B. Y., Hamledari, H., Shahi, A. & Zangeneh, P. (2017). Roles, benefits, and challenges of using UAVs for indoor smart construction applications. *Computing in Civil Engineering*, June 25–27, Seattle, WA.

Moeini, S., Oudjehane, A., Baker, T. & Hawkins, W. (2017). Application of an interrelated UAS–BIM system for construction progress monitoring, inspection and project management. *Project Management World Journal*, VI(VIII), 1–13.

Morgenthal, G. & Hallermann, N. (2014). Quality assessment of unmanned aerial vehicle (UAV) based visual inspection of structures. *Advanced Structural Engineering*, 17(3), 289–302.

Nex, F. & Remondino, F. (2014). UAV for 3D mapping applications: A review. *Applicable Geomatics*, 6(1), 115.

O'Neill, K. (2016). Spectacular aerial drone footage gives unique view of controlled hospital demolition. *Mirror*. Retrieved from: http://bit.ly/2xbAQaB

Park, J. & Cai, H. (2017). WBS-based dynamic multi-dimensional BIM database for total construction as-built documentation. *Automation Construction*, 77(Supplement C), 15–23.

Park, J., Kim, K. & Cho, Y. K. (2017). Framework of automated construction-safety monitoring using cloud-enabled BIM and BLE mobile tracking sensors. *Journal of Construction Engineering Management*, 143(2), 05016019-1-12.

Parramatta Advertiser. (2017). Drone captures demolition of Parramatta Stadium. *News Corp*. Retrieved from: http://bit.ly/2mqXAhE

Pritchard, L. (2015). Pointer avionics skyhunter406 – main. Retrieved from: http://www.skyhunter406.com.

Rumane, A. R. (2016). *Quality management in construction projects*. Boca Raton, FL: CRC Press.

Siebert, S. & Teizer, J. (2014). Mobile 3D mapping for surveying earthwork projects using an unmanned aerial vehicle (UAV) system. *Automation Construction*, 41, 1–14.

Stark, B. (2017). What drones may come: The future of unmanned flight approaches. *The Conversation Media Group*. Retrieved from: http://bit.ly/2w9lQ8l

Szentpeteri, K., Setiwan, T. R. & Ismanto, A. (2016). Drones (UAVs) in mining and exploration. An application example: Pit mapping and geological modelling. In *Proceedings of the MGEI 8th Annual Convention*, 05–06 October, Bandung, West Java, Indonesia, 45–49.

Taylor, C. (2017). Drones in construction predictions: Drones for demolition. *Uplift*. Retrieved from http://bit.ly/2xQPEIK

Tezel, A. & Aziz, Z. (2017). From conventional to IT based visual management: A conceptual discussion for lean construction. *Journal of Information Technology Construction*, 22, 220–246.

Trujillo, M. M., Darrah, M., Speransky, K., DeRoos, B. & Wathen, M. (2016). Optimized flight path for 3D mapping of an area with structures using a multirotor. In *Proceedings of the 2016 International Conference on Unmanned Aircraft Systems (ICUAS)*, 07–10 June, Arlington, VA, 905–910.

Vacanas, Y., Themistocleous, K., Agapiou, A. & Hadjimitsis, D. G. (2016). The combined use of building information modelling (BIM) and unmanned aerial vehicle (UAV) technologies for the 3D illustration of the progress of works in infrastructure construction projects. In: Themistocleous, K., Hadjimitsis, D. G., Michaelides, S. & Papadavid, G. (Eds.), *Proceedings of the Fourth International Conference on Remote Sensing and Geo-information of the environment*. Cyprus: SPIE, 96881Z, https://doi.org/10.1117/12.2252605

Wang, J., Sun, W., Shou, W., Wang, X., Wu, C., Chong, H.-Y., Liu, Y. & Sun, C. (2015). Integrating BIM and LiDAR for real-time construction quality control. *Journal of Intelligent Robot System*, 79(3), 417–432.

Wang, L., Chen, F. & Yin, H. (2016). Detecting and tracking vehicles in traffic by unmanned aerial vehicles. *Automation Construction*, 72, 294–308.

Wen, M.-C. & Kang, S.-C. (2014). Augmented reality and unmanned aerial vehicle assist in construction management. In: Issa, R. I. & Flood, I. (Eds.), *Computing in civil and building engineering*. Orlando, FL: American Society of Civil Engineers.

11 Gamification for sustainable construction

Introduction

Gamification is explained as the application of game design elements without making reference to gaming contexts or the application of elements of playing games like point scoring to other areas of action, typically as a way of marketing an online way of encouraging further service. Applications of games comprise simulations, gamification, and serious games in respect of the gamer's choice. As of 2010, gamification started becoming popular. It employs the technique of using game elements in a context that is unrelated to games in order to enhance loyalty, participation, and motivation of people. Several fields have implemented the use of gamification into their activities. Some of these fields include marketing, healthcare, and even education. In order to enhance users' engagement in construction, the implementation of gamification can be applied in many different ways. However, it is one thing to implement its use and another to always guarantee that such implementation in whatever form it can achieve the goal it was meant to. Hence, it is often necessary to accompany gamification with other available related techniques. Researchers have shown that including some social elements of gamification will further improve users' engagement. However, the greater the service that is being used and the longer the users interact with it, the better they gain individual experience of the possible service they can benefit from it, and thus become a professional in its application.

Process of developing gamification

Regarding the research conducted by Morschheuser et al. (2017), the gamification project development process involves the following eight phases:

- Testing and monitoring
- Design
- Project preparation
- Analysis

DOI: 10.1201/9781003179849-11

- Implementation
- Ideation
- Evaluation
- Testing and monitoring

Project preparation

This project stage at which gamification is projected implies becoming familiar with the specific problem that will be solved with the assistance of gamification. It also determines the best solution that is suggested to making measures on the success of a particular project. What to relate to in this process is to determine the project goals and ascertain the viability of the choice in going for gamification as the best option that is effective for achieving the objectives identified. Organisation that will make use of this process should draw up a project plan in with details that are precise, such as the deadlines, the duration of the scheduled work, the budget allocated, success criteria, and project teams to be involved (Burke, 2014; Morschheuser et al., 2017).

Analysis

This is the second stage of the process that involves the collection of data and necessary documents about the specific audience and the area of implementation of gamification. Ascertaining the goals can be done through surveys, interviews, observations, and group discussions according to Werbach and Hunter (2012) and Marache-Francisco and Brangier (2013). Morschheuser et al. (2017) explained that by the simple application of these methods, firms that engage in the usage of gamification will be able to recognize users' needs, their motives, and behavioural attitudes. Burke (2014) also emphasised the need for creating player personas, imaginary individuals with some common features that correspond with characters of a peculiar group of people, since it is helpful to avoid abstract discussions within the identified project group.

Ideation and design

This involves brainstorming and consolidating different ideas. At this stage, people come up with as many ideas as possible in order to find different design alternatives (Kapp, 2012; Deterding, 2015; Morschheuser et al., 2017). According to Masie (2015), aspects to consider when developing the design ideas are the following:

1 Every participant should be able to identify with the shaping and personalising personal gamification experience using his knowledge as a yardstick.

2 A player (participant) should experience minimal safe failure on the path to success.
3 Gamification elements can be a success no matter where they are applied – whether in groups or individually.

The idea has to be linked with the design. After the ideas have been collected, the next thing to do is to focus on establishing and developing designs that should be tested repeatedly to consider how effective they are towards achieving sustainability. This is followed by identifying the ways to make it work better and from there establish a concept to develop an update (Brito et al., 2015; Morschheuser et al., 2017).

Implementation of a design

This phase of the gamification development process includes all the information needed for a design implementation with outcome to create pilot (Fitz-Walter, 2015). Once the pilot is prepared, it is necessary to assess whether the design meets the objectives that it was meant to satisfy and from there move forward towards the goals identified.

Evaluation and testing

Helms et al. (2015) and Morschheuser et al. (2017) explained that the evaluation could be achieved through conducting interviews, regular or periodic testing, and surveys by putting into interest the way the users react to the number of various results to be obtained at the end of it all.

Monitoring

It is necessary to do continuous monitoring of a project in order to come up with good and improved results (Radoff, 2011).

Drivers and challenges of applying gamification

Globalisation and improvements in modern technology such as video games, mobile phones, and social media have enhanced the way people now view gamification. They now realise that it can be implemented in a variety of disciplines for achieving different objects and goals (Kim, 2015). However, in order to understand the ways by which gamification works, it is advantageous to have a basic idea of what it entails and what the benefits are of implementing gamification, as well as the challenges along with the criticisms that come with it.

Advantages of gamification

1 It increases user's involvement and engagement.
2 Gamification encourages collaboration, and the development of a culture of innovation, by delivering a range of ideas, and creating an ongoing relationship among users (participants).
3 Gamification identifies skills and makes for easier performance appraisals.

According to Chou (2016), there are some drivers of gamification. They are as follows:

1 **Epic meaning and calling:** Gamification users feel motivated that they are involved in something tangible by themselves and that the purpose coupled with the impact of their commitment is significant and recognised in a meaningful way.
2 **Development and accomplishment:** Gamification creates personal drivers that encourage people to progress and to develop new skills by achieving mastery in overcoming the challenges they faced.
3 **Ownership and possession:** The users are always motivated because they are participating in a process, coupled with the feeling of having something worthwhile that they could direct and control; and when a person experiences ownership over something, the person understandably becomes willing to improve what he or she possesses or owns.
4 **Social influence and relatedness:** This completes the drivers that express indirectly to all available and social elements that are reasonable and also motivates people, such as social feedback, competition, mentorship, and social acceptance or even envy.
5 **Scarcity and impatience:** The insatiable nature of people come to play with always wanting more, especially if the supply of such wants is limited, not readily available, or exclusive. For this reason, many games have tortured breaks or involved dynamics in terms of appointment and delivery.
6 **Unpredictability and curiosity:** By having practically little idea of what may happen within an unexpected time, users can stay motivated and excited about the next gaming experience.

Challenges for using gamification by:

1 Internal purchase-in problem, which explains that concept of utilising gamification for project significances, should be well annotated and accepted by the person in charge of management.
2 Before we can reach a successful outcome of a gamification experience, it can only be feasible when there is clear communication in which the users share mutual passions, interest, goals, engagement, and motivation.

3 Another challenge identified is when making systems or operations related to gamification, it has to be meaningful and the concept of defining what is meaningful rests on individual's perspective of the subject.

(Prakash & Rao, 2015)

Gamification for construction sustainability

The prominent way of bringing up affirmative environmental actions, such as green guides, newsletters, and posters all have little and also short-term success (Yen, 2015). Researchers highlight the growth of gamification in the construction industry as defined in the use of games to make sustainable construction networks for functioning construction practices (Kamal, 2013).

Froehlich (2015), Seaborn and Fels (2015), Dymek and Zackariasson (2016), and Mazur-Stommen and Farley (2016) all explained the need for sustainability in the construction industry. Gamified experiences should lead to a more competitive environment by comparing the outcome of each result. These results could be channelled in combating unnecessary material wastage and also using it to create awareness about the structures and concepts that are now employed in the construction industry. Szaky (2016) proposes that gamification can be applied to engage vernal generations to the issues that might arrive due to non-sustainability if they are disregarded. The researcher proposes educating them to implement environmental habits.

Relationship between gamification and sustainability

Gamification is the deliberate use of game concepts for a non-game task and background interaction (Robson et al., 2015). Concerning the principles of sustainability and gamification, Seaborn and Fels (2015) found that sustainability technologies are aimed at promoting and facilitating sustainable practices such as reducing the number of resources used, investing in recycling programmes and renewable forms of energy, and reusing content wherever possible (Seaborn, 2015). Therefore, as stated by other researchers (Robson et al., 2015), gamification may address business processes with the objective of changing or engaging workers, participants, or communities of interest in specific behaviours. Tourism and hospitality businesses may incorporate and use game elements to trigger other forms of behaviour. You can encourage visitors to reduce energy and water waste rates, use public transportation, or price facilities.

The key drivers of gamification remain in human behaviour's motivational drivers, according to Robson et al. (2015): reinforcements (extrinsic and intrinsic) and emotions. Such components motivate players (users) to learn and replicate the expected behaviour because most gamification schemes are competitive, while the game itself is an inherent reward for the player, even greater than conventional encouragement (Seaborn, 2015). Professionals and people involved should incorporate the goals of the companies into game mechanics, building the gamified interaction system that will inspire the behaviours of players or cause changes in behaviour (Robson et al., 2015).

Froehlich states that in many cases, businesses are using gamification to create a green or eco profile in order to increase the visibility of their brand image and promote new innovations rather than sustainable actions (Froehlich, 2015). Therefore, businesses must be mindful of the 'challenges posed by life's over-gamification.' It is not possible to gamify all the tasks in a business. It is also not recommended that everything is turned into a competition or that players plan to collect points for every stage of exploitation (Anderson & Rainie, 2012).

Gamifying construction

Operators and workers have risky jobs in this field. They do not have the luxury of making mistakes when practicing. A mistake can cost a couple of lives. What about workers without previous experience? Where are they going to learn and practice? Gamification comes in handy here. This creates a virtual environment in which workers can execute their work tasks without fear of causing harm to the real world. In an environment that is safe, they can make as many mistakes as they want and learn from their mistakes. As a result, the staff becomes more comfortable in doing their tasks, which ultimately leads to far fewer shop floor errors (Sayantani, 2017). The staffs are likely to better recall their errors because they seek immediate feedback and stop making mistakes.

Gamification encourages companies to bring out their staff's competitive streak. For example, organisations may award scores based on their behaviour and success. Then these scores are compared with those of other participants. When the participants try to compete with each other to improve their results, their performance will automatically be improved. It results in better activities on the shop floor over time (Sayantani, 2017). This method helps each employee recognise their areas of weakness and change in their work. Practice perfects their work skills. Apart from saving lives, minimised injuries save the company a great deal of money. It is time for the construction industry to understand gamification's long-term benefits and apply it to improve workplace safety.

Construction application

Employee learning probably offers the biggest opportunity in the construction business to reap the benefits of gamification. Because the industry, at least for on-site workers, tends to attract relatively young people, its employees may be especially receptive to game-based learning. Safety training, for example, is an area that requires continuous attention. However, it is also possible for staff to tune out. Framing security updates and drills in game scenarios where participants can win or lose ground by implementing safe or unsafe work practices is a way of making the system lively.

The construction industry is adapting to recent technologies that threaten the market. On construction sites, drones are becoming increasingly common. The construction industry has emerged as a key driver of the growing commercial drone industry, a trend that is expected to continue. Drones are a new technology and companies will need to train a professional drone operator. A range of drone training programs are already on the market for this reason. Such courses, without the risk of damaging costly aircraft, would teach the skills needed to fly. When technology advances and more expensive features become available, this will become increasingly important.

Simulations of game style can also help to prepare workers for positions in project management. Simulations in online training, set up as games, can test the decision-making and problem-solving skills of participants and allow them to see the potential consequences of various actions before granting them certain responsibilities in the field. They may also find games in which the participants can be rewarded based on meeting times, set budgets, and avoidance of work related accidents.

Finally, gaming has more impact on our lives than ever before in one way or another: the industry is booming. While the gamification movement in education has not been the quick success that could have been expected, there is still a function to play within a learning toolkit. With more jobs needed in industries especially construction, making education more enjoyable for both young people and adults alike will certainly be a winner.

Conclusion

Gamification can be described as a proactive tool for exercising the entire populace to have an attitude that is positive towards an environment that is suitable for all (Froehlich, 2015; Seaborn & Fels, 2015). Apart from this, game-related systems are able to provide other outcomes, like raising awareness of general environmental challenges and providing possible solutions. The construction industry is a complex industry and should be managed by related professionals by using gamification-related practices in executing construction for the sake of saving the environment. Therefore, gamification can be considered in solving some critical problems or challenges using game-like solutions. However, gamification is directed to assist in aspects that are beyond description, valuable, and important in context of environmental sustainability, but this can only be achieved when the right frame of mind is focused on making it work. When properly planned to cater for an identified aspect of work, it is expected to provide results.

References

Anderson, J. & Rainie, L. (2012). *The future of gamification.* Retrieved from: http://www. pewinternetorg/ 2012/05/18/the-future-of-gamification/

Brito, J., Vieira, V. & Duran, A. (2015). Towards a framework for gamification design on crowdsourcing systems: The G.A.M.E. Approach. In *Proceedings of the 12th International Conference on Information Technology – New Generations*, 13–15 April, LasVegas, NV, IEEE, 445–450.

Burke, B. (2014). Gamify: How gamification motivates people to do extraordinary things. Retrieved from: books.google.com.ng

Chou, Y. (2016). *Actionable gamification. Beyond points, barges and leader boards.* Fremont, CA: OctalysisMedia.

Deterding, S., (2015). The ambiguity of games: histories and discourses of a gameful world. In: Walz, S. P. & Deterding, S. (Eds.), *The gameful world: Approaches, issues, applications.* Cambridge, MA: MIT Press, 23–64.

Dymek, M. & Zackariasson, P. (2016). *The business of gamification: A critical analysis.* New York: Routledge.

Fitz-Walter, Z. (2015). *Achievement unlocked investigating the design of effective gamification experiences for mobile applications and devices.* Brisbane: Queensland University of Technology.

Froehlich, J. (2015). Gamifying green: Gamification and environmental sustainability. In: Walz, S. & Deterding, S. (Eds.), *The gameful world.* Cambridge, MA: MIT Press.

Helms, R. W., Barneveld, R. & Dalpiaz, F. (2015). *A method for the design of gamified trainings.* Singapore, AIS: PACIS.

Kamal, A. (2013). Green gamification: The apps, sites, and people that are going to save our llanet. Retrieved from: https://venturebeat.com/2013/01/30/green-gamification-the-apps

Kapp, K. M. (2012). The gamification of learning and instruction. *Game-based methods and strategies for training and education.* San Francisco: Pfeiffer.

Kim, B. (2015). Understanding gamification. *Library Technology Reports*, 51(2), 1.

Marache-Francisco, C., & Brangier, E. (2013). Gamification and human-machine interaction: a synthesis. *Le Travail Humain Journal Article*, 78(2), 165–189.

Masie, E. (2015). *The gamification game. Apply design, evidence and data to highlight the pros vs cons.* Chicago: MediaTex Publishing.

Mazur-Stommen, S. & Farley, K. (2016). *Games for grown ups: The role of gamification in climate change and sustainability.* Washington, DC: Indicia Consulting LLC.

Morschheuser, B., Werder, K., Hamari, J. & Abe, J. (2017). How to gamify? A method for designing gamification. In *Proceedings of the, on System Sciences (HICSS)*, 04–07 January, Hawaii, USA, 4–7.

Prakash, E. C. & Rao, M. (2015). *Transforming learning and IT management through gamification. International series of computer entertainment and media technology.* Switzerland: Springer International Publishing.

Radoff, J. (2011). *Game on: Energize your business with social media games.* Indianapolis, IN: Wiley Publishing.

Robson, K., Plangger, K., Kietzmann, J. H., McCarthy, I. & Pitt, L. (2015). Is it all a game? Understanding the principles of gamification. *Business Horizontal*, 58, 411–420.

Sayantani, B. (2017). How gamification can prevent accidents in construction sector. Retrieved from: http://news.elearninginside.com

Seaborn, K. (2015). Gamification in theory and action: A survey. *International Journal of Humanity Computation Study, 74,* 14–31. https://doi.org/10.1016/j.ijhcs. 2014.09.006

Seaborn, K. & Fels, D. I. (2015). Gamification in theory and action: A survey. *International Journal of Humanity Computation Study, 74,* 14–31.

Werbach, K., & Hunter, D. (2012). *For the win: How game thinking can revolutionize your business.* Wharton Digital Press.

Yen, M. (2015). *Eco-gamification: Game changer for sustainable real estate.* Retrieved from http://green-perspective.blogspot.fi/2015/09/eco-gamification-game-changerfor. html

12 Internet of things and sustainable construction

Introduction

Over the years, the construction sector has been regarded as a resource-intensive sector that consumes large chunks of global raw materials such as 25% of timber; about 16% of the water available on earth; and 40% of stone, gravel, and sand (Arena & De Rosa, 2003; Berardi, 2013; Shi, Chen & Shen, 2017). Also, the International Energy Agency (IEA, 2013) opined that the construction industry consumed more than 40% of the energy in the world. The resultant effects are a great deal of greenhouse gas emissions globally. In another vein, Tam and Tam (2008) maintained that the construction sector activities have generated a great deal of solid waste, wastewater, noise, significant dust levels, and smoke. Furthermore, if the situation is not addressed, it has been forecast that the energy consumption and greenhouse gas emissions of the construction sector would have risen by more than 50% by the year 2050 (Berardi, 2017). A common concept used for linking daily objects with an interactive network with the use of wireless communication mediums such as smart phones, radio frequency identification tags, and sensors is called the Internet of Things (IoT). This is applicable to many fields that include transportation, industries, and civic infrastructure.

In response to various challenges resulting from construction activities, sustainable construction has emerged towards mitigating the impacts. Darko and Chan (2016) defined sustainable construction as an efficient method towards putting into operation sustainable development principles in the construction sector. These include social, environmental, and economic sustainability all through the entire durational cycle of construction projects. The practice cored in sustainable construction can be understood as the best process of planning and integrating designs, maintaining, operating, renovating, converting, and demolishing existing edifices in an environmentally friendly, and healthy (social) approach, and by means of a resource-efficient (economic) process (United States Environmental Protection Agency [USEPA], 2016). Sustainable construction encompasses aspects of social, environmental, and economic ways of constructability. In actual fact, the introduction of sustainable construction will help to augment the traditional method of construction, thereby leading to effective resources and energy consumption, reduction, or elimination of greenhouse gas emissions, productivity, and human well-being.

DOI: 10.1201/9781003179849-12

The need to create more efficiency in the ways we harness and maximise opportunities in our daily settings gave birth to the IoT. In 1999 during a presentation on supply-chain management, Kevin Asthon came up with the phrase 'IoT'. The expression of the word 'things' is believed by Asthon to be that the way humans interact with the physical environment needs serious re-evaluation due to the advancements in technology such as the Internet, computing, and the rate of data-generation by smart devices (Yu, 2014; Buyya & Dastjerdi, 2016). The 'IoT' paradigms definitions can be best understood using both the technological and application of the IoTs system; the application is the human-oriented vision while the technological comprises semantic-oriented vision, internet-oriented vision, and things-oriented vision (Jia et al., 2019). The IoT unites a system and creates ease of delivery for humanity. Sustainable construction is fast growing in both developed and developing countries of the world such as the United Kingdom, Canada, the United States, Australia, and Nigeria, just to mention a few. The IoT is becoming popular in terms of timely completion, organisation of routine schedules, and ease of works in various sectors such as government agencies, manufacturing, and smart buildings.

Internet of things

Business Intelligence (2015) forecasted that the IoT would experience significant growth from 5-million devices from 2004 to 35-million devices by 2018. Ortutay (2015) agreed that the IBM-USA would spend billions of dollars on the research relating to the IoT and how it can be translated into industrial usage. In the same vein, Wang et al. (2015) reported that the United Kingdom had employed the usage of IoTs on a five-million GBP project. Also, Yu (2014) affirmed that Singapore had announced the usage of IoTs in surveillance cameras and other sensory devices, thus enabling the country towards becoming a smart nation. The increase in the usage of IoTs in various organisations is becoming evident from various available works of literature online. Diverse sectors are looking at maximising profits and reducing costs using smart technology like the IoT. Smart technology like the IoTs is leading the universe into a ubiquitous market place. The IoT is a leading line of technology that encompasses how man and machine can interact; the IoTs bring things in reality, with sensors that are plugged into the Internet through wire or wireless types of connections.

The IoTs architecture

Buyya and Dastjerdi (2016) opined that the building blocks of IoTs have been around for years but without a unified network of how it should be operated. Some of the building blocks of IoTs are remote service, the communication network, sensory devices, and context-aware processing of events. He opined that the IoT is a unified network that brings human beings and smart objects together; the human can be responsible for the operation of the smart object or the smart object can operate automatically, thereby enabling a system in which humans and smart objects ubiquitously communicate with each other. Buyya and Dastjerdi (2016) also added that an ideal IoT architecture must be capable of operating flawlessly with its components and should be able to link together the virtual and physical realms. The IoT's architecture is composed of a dashboard or web portal, management of the application programming interface (API), device management, event processing and analytics, management of resources, service repository and discovery, and a message broker along with a communication layer, security and privacy enforcement, and identification, authorisation and access control. The entire above are achieved using devices, sensors, and human operators (Buyya & Dastjerdi, 2016).

The communication process of IoT

Castellani et al. (2010) opined that the communication and network of IoT is an aggregation of a mobile adhoc network (MANET), and the mobile network such as 4G-Lite, 3G, a wireless sensor network (WSN), and a wireless local area network (WLAN). Rimal, Choi and Lumb (2009) classified the communication layers of IoT into the application phase, transport phase, network phase, and physical and link phase (layer).

Sustainable construction

Green building, sustainable building, and green construction terminologies that have been employed by different researchers are all terms still within the scope of sustainable construction. The core aspects of sustainable construction have been on how the technical, social, environmental, and economical aspects of the world can be harnessed efficiently and effectively in the present generation without compromising the future generation. In response to this, various sustainability technologies have been employed such as water efficiency, energy efficiency, indoor quality environmental enhancement, materials and resources efficiency, and control systems (Lee et al., 2007; Zhang et al., 2011; Roufechaei et al., 2014; Chen et al., 2015; Koebel et al., 2015; Ahmad et al., 2016). Research have been carried out already by several scholars that addressed the concept behind sustainability is in construction, with some identifying the barriers, drivers, project delivery attributes, critical success factors, risks, and shareholders. Some of the benefits of sustainable construction are reduction in water and energy consumption, better quality of physical and psychological health, mental focus, occupant comfort, and increase in productivity (Fisk & Rosenfeld, 1997; Fuerst & McAllister, 2009; Loftness et al., 2009; Bluyssen, Aries & van Dommelen, 2011; Diamond et al., 2013; Boubekri et al., 2014). The social aspect of sustainability has to do with the people, the technical has to do with the professionals involved in the delivery of sustainable construction, and the environment has to do with our environment, while the economical has to do with whole life cycle of the overall cost of sustainable construction.

Internet of things for sustainable construction

Its application is becoming popular in different fields such as construction, man-ufacturing, and aviation. How the IoT can be applied to achieve sustainable con-struction is discussed in this section. The section discusses areas where IoT has been applied in buildings with a view to applying the concepts towards achieving sustainable construction as summarised in Figure 12.1.

Priority on a control system

The control system in sustainability has to do with the management of occupant building environment, the energy conservation, luminance and thermal comfort, and internal air quality (Dounis & Caraiscos, 2009). Ahmad et al. (2016) iden-tified seven control systems for sustainable building projects: occupancy sensor, heating, ventilation and air conditioning (HVAC) control system, audio-visual control, intercoms, security control, and shading control. Some of the goals of these systems are to ensure a unified comfort for the inhabitant of sustainable building projects, for example, the HVAC is to maintain energy use and thus provide comfort for the occupants.

Priority on efficient use of energy

One of the core aspects of sustainability in construction is to achieve energy efficiency for construction, operation, and maintenance. Energy efficiency is the lower limit of required energy to power goods and services. There are numerous technologies that been applied to energy-efficient building construction. Zhang et al. (2011) pointed out that housing developers can achieve energy efficiency by

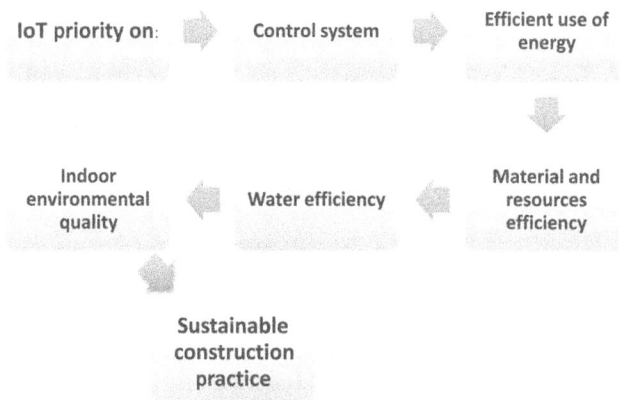

Figure 12.1 Internet of things application in construction industry.

using solar heating technology and the low emissivity insulation windows. Also, Chen et al. (2015) acknowledged that the use of technological advancements such as natural light and ventilation, shading devices, and building orientation optimisation can help towards the reduction of the energy budget and better energy efficiency.

Priority on indoor environmental quality

This has to do with how the building occupants will be provided with effective indoor environmental quality. The technology that has been applied is thermal control, thermal performance, and ample ventilation for the pollutants (Zhang et al., 2011). In addition, the use of technology such as thermal insulation, the application of ceiling heights and the enhancement of chimneys for stack ventilation are considered important towards maintaining a comfort temperature for inhabitant (Ahmad et al., 2016).

The priority on water efficiency

As part of conserving resources usage, water efficiency is a core aspect of sustainable construction. Ahmad et al. (2016) identified two technologies that are used for water conservation in a sustainable building: efficient water appliances and rainwater harvesting. In the same vein, Zhang et al. (2011) indicated that in order to experience low-carbon communities and water efficiency, sustainable building must employ decentralised rainwater, grey water systems, and water-saving appliance.

The priority on material and resources efficiency

It is believed that materials and resources are scarce and non-renewable respectively. Researchers are on the lookout towards providing solutions to the scarce materials and non-renewable resources in order to achieve sustainable construction. Roufechaei et al. (2014) opined that HVAC system can be well suited if environmentally friendly materials are employed. Zhang et al. (2011) indicated that the use of underground space development material can also help the HVAC system. There is continuous usage of various materials and conservative resources towards achieving sustainable construction.

Other building area relating to the study

Yang, Wang and Zhang (2015) in China used mobile phones and tablets in occupant localisation for hospital department route direction using related data of personal information and iBeacon ID. This iBeacon helps to send information to the server when the patient enters the hospital, and the server sends back information to the patient on the shortest route to her/his ward using 4G, 3G, and Wi-Fi. Alletto et al. (2015) in Italy used processing centres and multimedia walls in occupant localisation for artwork information using related data of users' location and all sorts of content information of each network. The users are able to relate with the systems by determining the artwork for his/her perusal. Lee et al. (2017) in Korea used intelligent mobile robots in general localisation using related data of Wi-Fi signals. The IoT application helps to determine assets and resource tracking.

Pan et al. (2015) in the United States used cloud computing and smartphone GPS-enabled in tracking assets in buildings for recording physical equipment by using related data of appliance electricity. Viswanath et al. (2016) in Singapore applied the use of a multi-purpose node, BLE sensor, smart plug, and smart gateway for the achievement of an energy control system in a residential building. The function of the systems is dependent on dynamic pricing information. The application allows users to select low load-consuming tasks within the system. Depuru et al. (2011) in the United States used smart meters in the power grid for general building types and they were able to determine real-time energy consumption. D'Elia et al. (2010) studied business building; the application was how to achieve building maintenance for end-users. The key technology of IoT that was applied was focused on ontology design and application development. The study helps to report temperature range and humidity; if the temperature is beyond the range, a message is generated and end-users are notified of the location of the faulty equipment. Within the system, it details the corrective maintenance to be employed to the personnel and occupants.

Srinivasan et al. (2017) applied the use of the distributed acoustic platform to realise preventive maintenance for the HVAC system in educational buildings in the United States. The data was sourced from audio signals which are transmitted via the Internet and processed with the HVAC system. The sensors were used to store information of CO_2, temperature, light intensity, and humidity emanating from the system. Streather in Streather (2016) explained that the United Kingdom applied sensors and cloud platform for analysis in soft facility management and hard facility management in commercial buildings. In soft facility management, sensors are installed under desks in order to determine open spaces for booking.

Kelly, Suryadevara and Mukhopadhyay (2013) in New Zealand used Zigbee WSN, web interface at the user end in the monitoring of domestic housing conditions. The related data that was sourced were attributes of environmental conditions, current, hot water, and voltage of appliances. The main purpose of this system is to establish the interconnection reference of IPv6 (Internet Protocol version 6) and WSN (Wireless Sensor Network) to accomplish low costs while still maintaining the integration of IoT with the functional unit of home monitoring systems.

Bashir and Gill (2016) in Australia applied the use of Hadoop in incorporating big data to analyse the indoor environment for a general type of building. The related data that was simulated were luminosity, oxygen level, and hazardous gases. Using the Hadoop, they were able to control the system if any of the simulated data was out of range, thereby enabling them to maintain a comfortable pre-determined range. The United States applied the use of IoT in industrial office buildings for intelligent building solutions i.e. energy saving and comfort. The data used was indoor temperature. This makes use of smart phones that allowed office occupants to send complaints to any of the analytic engines, thereby enabling the feedback to maintain optimal temperatures within the HVAC system.

Demirkan (2013) in the United States applied cloud services, big data, BFID/barcode and web 2.0/3.0 in health care for healthcare services for easy of portability and ease of deployment. Li et al. (2011) in Canada used home surveillance systems and body sensors in the determination of pervasive healthcare while a neighbourhood watch in residential buildings using related data of body sensor readings from a personal healthcare information system. Piscitello et al. (2015) in Italy used virtual sensors, DangerCore, and 4g, Wi-Fi, SMS, and mobile devices in determining occupant safety and emergency management using related data of feedback of users such as noise. Patel and Panchal (2018) used smart gateway, Arduino microcontroller, and PIR sensors in India for reservation facility automated parking system and smart meeting space (area) with real-time room occupancy status.

Kovac et al. (2015) in the Czech Republic used Raspberry, a simple network management protocol (SNMP), openHab, and hypertext transfer protocol (HTTP) to remotely monitor the automation system in different homes. Interconnected communication systems can be remotely monitored by occupants. Wang, Fu and Yang (2017) in China used idataBox, sensing layers in determining an early warning system and structural health monitoring for a six-storey building. The related data used were steel stress and pressure. Salikhov et al. (2016) in Russia used door sensors, Raspberry Pi, human trackers, and sensor tags in Microservice-based IoT systems using related data such as temperature, occupancy, pressure, door status, luminosity, and humidity to determine occupants' comfort. Wei and Li (2011) in China applied the usage of sensor networks in general types of buildings in the energy management system and energy consumption. The related data used are sensors from HVAC, power distribution systems, environment information, lighting systems, and plumbing. The system contained three layers with each having a subsystem; each layer manages energy consumption, fault analysis, and monitoring of equipment.

From above, the major application area of IoT in buildings has been for localisation, facility management, energy management, indoor comfort, and safety and security. Most of the related building areas have been centred on HVAC, lighting, security alarm system, auxiliary facilities, and the structure. Also, the layers employed range between the perception layer, application layer, and network. In essence, some of the IoT's third applications used are cloud computing, encryption, Bluetooth, long-term evolution (LTE), radio-frequency identification (RFID), ambient sensors, smart meters, and the like. Sustainable construction can be integrated with IoTs towards achieving energy efficiency, material and resources efficiency, water efficiency, environmental quality, and control system. Sustainable construction is cost-intensive; however, the IoT can serve as a signal whenever resources are going beyond the scope of the project. IoT can bring stakeholder of sustainable construction together seamlessly, thereby enhancing the flow of information. Effectiveness and efficiency are best served when we integrate IoTs with sustainable construction toward achieving the quadruple of sustainability which is economic, social, environment, and technical sustainability.

Internet of things' roles in sustainability

The IoT is an important technological method that is expressed towards information-communication linked with the smart cities movement, especially in the domain of sustainable development. It is also important that a link be drawn between smart cities and the notion of sustainability because the usage of the IoT is deeply rooted in the context of smart cities. Conceptually, smart cities can be said to be a part that is working in the direction of sustainability in the sense that smart cities seek to make the most of the benefits derivable by most people with the least cost and impacts which spells out clearly the very goal of aiming towards sustainability. The American Planning Association (APA) Smart Cities and Sustainability Initiative regards smart cities as an extension and a vital part heading towards sustainability (APA, 2015).

Roles of internet of things for sustainability

Collaboration models and incentives alignment

1 Sustainability objectives should be prioritised at the design face of the project through its provision of structural incentives.
2 Construction models and scales are facilitated via infrastructural solutions.
3 The data governance term for ownerships is established early enough to also include usage, privacy, and sharing as a unique pillar of the partnership.
4 Legal frameworks are simplified, procurement processes are accelerated and experts are engaged to maximise the speed of IoT's deployment and in turn political cycle risks are reduced.

Models for business and investment

1 Flexibility is exercised in the designing and execution of the construction project.
2 Mutual benefits are unlocked through the development of cross-industry solutions and also new monetisation models are enabled.
3 Alternative funding sources (e.g. institutional investors) are attracted by demand consolidation and bundling.

Impact measurement

1 In response to new generational demand for sustainable construction, a sustainability awareness culture would have to be put in place. Brand reputation would have to also be improved upon.

2 Based on UN Sustainable Development Goals, a framework can be adopted in order to evaluate potential impacts and to measure results.
3 Potential sustainable development goals are identified and targets addressable by the IoT are identified and incorporated into the design.

Drawbacks with IoT in the actualization of sustainable construction

Some of the major challenges encountered through the use of the IoT technology in enhancing sustainable construction are listed below. Some relevant approaches have also been recommended alongside to overcome these challenges.

Lack of incentive

As suggested by Greenfield, an important feature of sustainable construction is that enterprises rather than government or institutions are playing a vital role in the invention of technical systems and paradigms for the concept of sustainable construction. Through the reliance on commercial sectors, there is a reflection of the importance of business factors in the usage and application of IoT technology on a large scale. However, this characteristic of sustainable construction might lead to the problem of a lack of incentives for building a sustainable sensor network owing to its limited profit magnitude. In order to overcome this problem, one can incorporate sustainable features that have other functions embedded as seen in the application of smart homes. Significant contributions to the business vision of IoT in the field of sustainable construction can be seen in the collaboration of technological endeavours and real estate developers.

Various sectors of the city such as planning professionals, technical firms, government and real estate developers have to come together to build incentives required for the IoT on a large scale. There are numerous ways to address the application and collaboration of IoT technologies in construction sustainability. This is derived from existing examples. In the example of Padova, Italy, the public and private sectors made a contribution to the Padova Smart City project. A feasibility analysis was conducted by the University of Padova as well as data processing and the University provided financial support for the project. A firm called Patavina IoT was engaged to carry out the task in the field of sustainability. The technical core of the project as well as the software was designed by technologists.

Span

The geographical range that the network covers and the density of the sensor gadgets is what span covers or concerns itself with. Financial feasibility, policy motive, and physical operation are the three major factors that the span of a network sensor is dependent upon. The design and running of the network system, the maintenance and development, and the selection of instalment locations based on vital needs are all examples of physical operation.

In the issue of applying sensor networks in an advanced sustainability infrastructure, the location and coverage of different devices can often be a city-wide matter, which requires the coming together of policy makers, technicians, and town planners. A higher level of precision can be attained when a sensor network becomes denser. The larger the network scales for the infrastructure, the higher the cost of building the infrastructure. Carry on here in a design of IoT system to enhance sustainable construction; the span of the sensor networks is an important factor to consider.

Physical forms of cities and buildings can be influenced or impacted by the specific location of instalments. Two factors that can affect the construction and renovation cost of the infrastructure are decisions as to whether to have embedded devices within new infrastructure or to add external devices to existing infrastructure. The coverage and strength of data detection within a certain sensor network is determined by the span. In a wider planning phase of the IoT projects, span can be said to involve the range of data required to assess and demonstrate requirements. Two issues are closely associated with the needs and motives of building an IoT system for sustainable construction, namely the kind of IoT system deployed and the kind of tracked data. Very crucial issues to discuss during the initial phase of any sustainable construction project are decisions on where to install, the question of what to install and what to detect.

Tolerance to fault

Another vital issue that is considered is the level of resilience of the sensor network system to system failure. That is the ability of the system to be able to continue functioning reasonably well even upon the detection of a fault. An example would be the sensor system sensor networks at the civic level which are often not capable of carrying out security protection. This now creates room for eavesdropping on vital information. Data security would thus have to be tagged as a vital issue considering the low information capacity of sensor devices. According to Adam Greenfield, optimization questioning is another problem related to fault tolerance. According to Greenfield, a seamless user experience is the major goal of sustainable construction; however, a major drawback of this is the ability to locate problem areas when they occur.

Ownership of data

The IoT for city administrators is emphasised by most of the literature as related to the subject matter in terms of optimised management. However, the issue of the ownership of data derived from sensors arises. A common feature of the smart cities movement is the administrator-centric mode which neglects the fact that a great deal of data is attributed to the end users or citizens. When there is no public access to data, citizens begin to experience difficulties when trying to access the benefits of a sensor network. This results in difficulty in sourcing funds and public support as well.

A more determined effort is thus required to build a link between the public and professionals in order to allow for active participation within the smart cities movement. This is in response to the ownership concerns of data collected from the sensor networks.

Adverse effect

Sustainable construction is improved through the use of the sensor network. The rebounding effect is another issue in which the increased efficiency of energy use will result in a corresponding increase in the amount of energy used. Similarly, as with the previous challenges, collaboration among different fields would help reduce the adverse effect of this.

Conclusion

In recent trends, it can be observed that the IoT is beginning to penetrate the building industry. The benefits and limitations of the IoT are being explored both by researchers and practitioners in a bid to fast track its actual implementation. For instance, it can be seen that companies comprising IBM and Intel are already launching their products of sustainable constructions to the world (Watson et al., 2018), illustrating the competitive advantage and future tendency of the IoT. Therefore, it is paramount for moves to be made on how to fully incorporate the IoT in sustainable construction so as to enhance the system itself. Despite the existence of IoT-based sustainable construction, there is a lack of understanding, review, and analysis on the application of IoT in the general fields for the development of buildings. Furthermore, with increasing interest in interdisciplinary research, researchers may consider an analytical review as a starting point for research in the area of construction and architecture engineering for sustainable construction. It is also valuable to understand that the technical needs and potential application areas required of the project should be well understood even if the IoT sector is goaded in order for the industry to experience enhancing different dimensions and also improve upon the development of sustainable construction.

References

Ahmad, T., Thaheem, M. J. & Anwar, A. (2016). Developing a green-building design approach by selective use of systems and techniques. *Architectural Engineering and Design Management*, 12(1), 29–50.

Alletto, S., Cucchiara, R., Del Fiore, G., Mainetti, L., Mighali, V., Patrono, L. & Serra, G. (2015). An indoor location-aware system for an IoT-based smart museum. *IEEE Internet of Things Journal*, 3(2), 244–253.

American Planning Association (APA). (2015). APA *policy guide for planning for sustainability.* Retrieved from: https://www.planning.org/policy/guides/adopted/sustainability.html

Arena, A. P. & De Rosa, C. (2003). Life cycle assessment of energy and environmental implications of the implementation of conservation technologies in school buildings in Mendoza. *Argentina. Building and Environment*, 38(2), 359–368.

Bashir, M. R. & Gill, A. Q. (2016, December). Towards an IoT big data analytics frame-work: Smart buildings systems. In: *Proceedings of the 2016 IEEE 18th International Conference on High Performance Computing and Communications; IEEE 14th International Conference on Smart City; IEEE 2nd International Conference on Data Science and Systems (HPCC/SmartCity/DSS)* (pp. 1325–1332).

Berardi, U. (2013). Clarifying the new interpretations of the concept of sustainable build-ing. *Sustainable Cities and Society*, 8, 72–78.

Berardi, U. (2017). A cross-country comparison of the building energy consumptions and their trends. *Resources, Conservation and Recycling*, 123, 230–241.

Business Intelligence (BI). (2015). *Research for the digital age*. Retrieved from: https://intel-ligence.businessinsider.com/

Bluyssen, P. M., Aries, M. & van Dommelen, P. (2011). Comfort of workers in office build-ings: The European HOPE project. *Building and Environment*, 46(1), 280–288.

Boubekri, M., Cheung, I. N., Reid, K. J., Wang, C. H. & Zee, P. C. (2014). Impact of windows and daylight exposure on overall health and sleep quality of office workers: A case-control pilot study. *Journal of Clinical Sleep Medicine*, 10(06), 603–611.

Buyya, R. & Dastjerdi, A. V. (Eds.). (2016). *Internet of things: Principles and paradigms*. Elsevier.

Castellani, A. P., Bui, N., Casari, P., Rossi, M., Shelby, Z. & Zorzi, M. (2010). Architecture and protocols for the internet of things: A case study. In: *Proceedings of the2010 8th IEEE International Conference on Pervasive Computing and Communications Workshops (PERCOM Workshops)*, 29 March–02 April, Mannheim, 678–683.

Chen, X., Yang, H. & Lu, L. (2015). A comprehensive review on passive design approaches in green building rating tools. *Renewable and Sustainable Energy Reviews*, 50, 1425–1436.

Darko, A. & Chan, A. P. (2016). Critical analysis of green building research trend in construction journals. *Habitat International*, 57, 53–63.

D'Elia, A., Roffia, L., Zamagni, G., Vergari, F., Bellavista, P., Toninelli, A. & Mattarozzi, S. (2010, June). Smart applications for the maintenance of large buildings: How to achieve ontology-based interoperability at the information level. In: *Proceedings of the IEEE Symposium on Computers and Communications*, 22-25 June, Riccione, Italy, 1077–1082.

Demirkan, H. (2013). A smart healthcare systems framework. *It Professional*, 15(5), 38–45.

Depuru, S. S. S. R., Wang, L., Devabhaktuni, V. & Gudi, N. (2011, March). Smart meters for power grid — Challenges, issues, advantages and status. In: *Proceedings of the 2011 IEEE/PES Power Systems Conference and Exposition, 20-23 March*, Phoenix, AZ, 1–7

Diamond, R., Ye, Q., Feng, W., Yan, T., Mao, H., Li, Y. & Wang, J. (2013). Sustainable building in China — A green leap forward? *Buildings*, 3(3), 639–658.

Dounis, A. I. & Caraiscos, C. (2009). Advanced control systems engineering for energy and comfort management in a building environment – A review. *Renewable and Sustainable Energy Reviews*, 13(6–7), 1246–1261.

Fisk, W. J. & Rosenfeld, A. H. (1997). Estimates of improved productivity and health from better indoor environments. *Indoor Air*, 7(3), 158–172.

Fuerst, F. & McAllister, P. (2009). An investigation of the effect of eco-labeling on office occupancy rates. *Journal of Sustainable Real Estate*, 1(1), 49–64.

International Energy Agency (IEA). (2013). Modernising building energy codes. *Policy Pathway*. Retrieved from https://www.iea.org/reports/policy-pathway-modernising-building-energy-codes-2013

Jia, M., Komeily, A., Wang, Y. & Srinivasan, R. S. (2019). Adopting internet of things for the development of smart buildings: A review of enabling technologies and applica-tions. *Automation in Construction*, 101, 111–126.

Kelly, S. D. T., Suryadevara, N. K. & Mukhopadhyay, S. C. (2013). Towards the implementation of IoT for environmental condition monitoring in homes. *IEEE Sensors Journal*, 13(10), 3846–3853.

Koebel, C. T., McCoy, A. P., Sanderford, A. R., Franck, C. T. & Keefe, M. J. (2015). Diffusion of green building technologies in new housing construction. *Energy and Buildings*, 97, 175–185.

Kovac, D., Hosek, J., Masek, P. & Stusek, M. (2015). Keeping eyes on your home: Open-source network monitoring center for mobile devices. In *Proceedings of the 2015 38th International Conference on Telecommunications and Signal Processing (TSP)*, 9th July 2015, 612–616.

Lee, S. K., Yoon, Y. J. & Kim, J. W. (2007). A study on making a long-term improvement in the national energy efficiency and GHG control plans by the AHP approach. *Energy Policy*, 35(5), 2862–2868.

Lee, S., Lee, N., Ahn, J., Kim, J., Moon, B., Jung, S. H. & Han, D. (2017). Construction of an indoor positioning system for home IoT applications. In *Proceedings of the 2017 IEEE International Conference on Communications (ICC)*, Paris, pp. 1–7. 10.1109/ICC.2017.799159

Li, X., Lu, R., Liang, X., Shen, X., Chen, J. & Lin, X. (2011). Smart community: An internet of things application. *IEEE Communications Magazine*, 49(11), 68–75.

Loftness, V., Aziz, A., Choi, J., Kampschroer, K., Powell, K., Atkinson, M. & Heerwagen, J. (2009). The value of post-occupancy evaluation for building occupants and facility managers. *Intelligent Buildings International*, 1(4), 249–268.

Ortutay, B. (2015). IBM to invest $3-billion in new 'Internet of Things' unit. *Globe and Mail*. Retrieved from: https://www.google.com/amp/s/www.thejakartpost.com

Pan, J., Jain, R., Paul, S., Vu, T., Saifullah, A. & Sha, M. (2015). An internet of things framework for smart energy in buildings: Designs, prototype, and experiments. *IEEE Internet of Things Journal*, 2(6), 527–537.

Patel, J. & Panchal, G. (2018). An IoT-based portable smart meeting space with real-time room occupancy. In Y.-C. Hu, S. Tiwari, K.K. Mishra & M.C. Trivedi (Eds.), *Intelligent communication and computational technologies*. Singapore: Springer, 35–42.

Piscitello, A., Paduano, F., Nacci, A. A., Noferi, D., Santambrogio, M. D. & Sciuto, D. (2015). Danger-system: Exploring new ways to manage occupants' safety in smart building. In *Proceedings of the 2015 IEEE 2nd World Forum on Internet of Things (WF-IoT)*, December 2015, 675–680. http://doi.org/10.1109/WF-IoT.2015.7389135

Rimal, B. P., Choi, E. & Lumb, I. (2009). A taxonomy and survey of cloud computing systems. In *Proceedings of the 2009 Fifth International Joint Conference on INC, IMS and IDC*, August, 44–51. https://doi.org/10.1109/NCM.2009.218

Roufechaei, K. M., Bakar, A. H. A. & Tabassi, A. A. (2014). Energy-efficient design for sustainable housing development. *Journal of Cleaner Production*, 65, 380–388.

Shi, Q., Chen, J. & Shen, L. (2017). Driving factors of the changes in the carbon emissions in the Chinese construction industry. *Journal of Cleaner Production*, 166, 615–627.

Salikhov, D., Khanda, K., Gusmanov, K. & Mazzara, M. (2016). Microservice-based IoT for smart buildings. Innopolis University, Russia. Retrieved from: http://www.researchgate.net/publication/309513228

Srinivasan, R., Islam, M. T., Islam, B., Wang, Z., Sookoor, T., Gnawali, O. & Nirjon, S. (2017). Preventive maintenance of centralized HVAC systems: Use of acoustic sensors, feature extraction, and unsupervised learning. In *Proceedings of the 15th IBPSA Conference*, 2017, 2518–2524.

Streather T., (2016). Internet of Things: digital disruption in facilities management. Retrieved from https://www.avnet.com

Tam, V. W. & Tam, C. M. (2008). Waste reduction through incentives: A case study. *Building Research & Information*, 36(1), 37–43.

United States Environmental Protection Agency (USEPA). (2016). *Definition of green building*. Retrieved from https://archive.epa.gov/greenbuilding

Viswanath, S. K., Yuen, C., Tushar, W., Li, W. T., Wen, C. K., Hu, K. & Liu, X. (2016). System design of the internet of things for residential smart grid. *IEEE Wireless Communications*, 23(5), 90–98.

Wang, F., Hu, L., Zhou, J. & Zhao, K. (2015). A survey from the perspective of evolutionary process in the internet of things. *International Journal of Distributed Sensor Networks*, 11(3), 462752.

Wang, J., Fu, Y. & Yang, X. (2017). An integrated system for building structural health monitoring and early warning based on an internet of things approach. *International Journal of Distributed Sensor Networks*, 13(1), 1550147716689101.

Watson, T., Boyes, H., Hallaq, B. & Cunnigham, J., (2018). The industrial internet of things (IIoT): An analysis framework. *Computer Industry*, 101, 1–12.

Wei, C. & Li, Y. (2011). Design of energy consumption monitoring and energy-saving management system of intelligent building based on the Internet of Things. In *Proceedings of the 2011 International Conference on Electronics, Communications and Control (ICECC)*, Ningbo, China, pp. 3650–3652. 10.1109/ICECC.2011.6066758

Yang, J., Wang, Z. & Zhang, X. (2015). An ibeacon-based indoor positioning systems for hospitals. *International Journal of Smart Home*, 9(7), 161–168.

Yu, E. (2014). *Singapore unveils plan in push to become smart nation*. Retrieved from: ZDNet: http://www. zdnet com/sg/Singapore-unveils-plan-in-push-to-become-smartnation-7000030573

Zhang, X., Platten, A. & Shen, L. (2011). Green property development practice in China: Costs and barriers. *Building and Environment*, 46(11), 2153–2160.

13 Machine learning for sustainable construction

Introduction

The evolution of machine learning, unlike in the past, along with its new ability to perform and apply difficult mathematical skills on complex and big data, repeatedly and faster is a new development with the introduction of modern computing technologies. Industry and organizations, e.g. financial institutions, transportation, oil and gas, health care, government, construction, retail, and security have been able to work efficiently and effectively with the introduction of machine learning that gives them the edge over their competitors. The schematic procedure of computer acceptability of data and the ability to process and produce output with less human intervention is said to be a subset of artificial intelligence. Machine learning, according to McAfee et al. (2012), understands data structure and ensures theoretical distribution of the understood data; even in the absence of a theory of a structure, its ability to use the computer to probe data for structure has developed machine learning.

Sustainable construction

According to Wikipedia and the United Nations Environmental Programme (UNEP), the current state of increased construction activities and infrastructural development may likely weaken and destroy the natural vegetation and wildlife if not properly addressed because construction activities have taken over half of humans' natural resources requirement. Sustainable construction is said to have emerged in the 1970s when the oil crisis was at its peak, and citizens felt the need for conservation of energy and the search for alternatives. It was reported that the need for sustainability was at its peak when there was a scarcity of other natural resources such as water.

It is no longer novel being 'green'; it is of utmost importance to uphold what is right for the environment. The act of sustaining the built environment produces high market values and dedication and commitment of the employees because a standard has been set when it comes to green construction on sustainability (Novotny, 2019). Sustainable construction focuses on upholding the current requirement (in construction) for the adequate provision of basic amenities and

DOI: 10.1201/9781003179849-13

infrastructural facilities without depriving the coming generation of their need for the provision of basic amenities. Sustainable construction can have social, environmental, technical, and economic impacts.

1 The social impact: This involves obeying the laid-down rules and regulations governing the construction industry throughout the building stages in order to provide a habitable living and good working environment for all the stakeholders involved.
2 The environmental impact: This governs the obligations to be followed by aiming towards a non-toxic environment.
3 The economic impact: Sustainable construction in this context involves market expansion, the re-using of waste, the preservation of materials, and the adaptability of buildings to changes that allow for high productivity and benefit.
4 The technical impact: This involves putting in effort that is infused with the right innovative ideas to make the work better than what it was.

Benefits of sustainable construction

As noted earlier in previous chapters, Wikipedia explains the benefits of sustainable construction as lower construction cost by high investment cost at the pre- and post-contract stages of the project. The introduction of technologies among competitive contractors would definitely lead to lower costs of construction. Other benefits include environmental protection, promoting sustainability, and market expansion. There are so many ways in which sustainability in construction can be achieved, e.g. prefabrication is the process of pre-forming construction components off site and bringing them to site for assembling with the benefit of producing elements/components to precision and specification without wastage of the material. Building information modelling (BIM) are models that help in the accurate calculation of the required materials that assists in the reduction of wastages and improves the building construction process. It can also help in improving the quality of the construction process efficiency with all the resources that go into the building construction (Novotny, 2019). This and many more is summarized in Figure 13.1 and Table 13.1.

Machine learning for sustainable construction

The choice of a new area in construction and the project monitoring of smart buildings are tasks being gradually unlocked with the implementation of large data in the construction industry used in tackling and assessing the success factor of a contract at the pre- and post-contract stages (Valpeters, Kireev & Ivanov, 2018). D'Amico et al. (2018) stated that there is a need for adequate data that serves as the only barrier in the effective utilization of machine learning. In understanding the impact of environmental building structures and with the

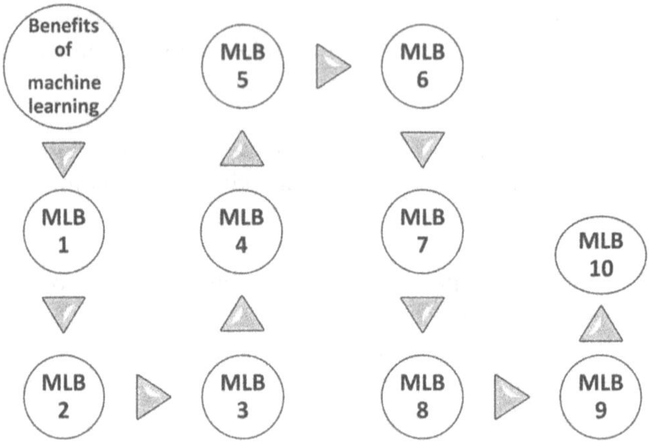

Figure 13.1 Benefits of machine learning in construction.

current state of machine learning, there is every probability that it will be implemented successfully for sustainable construction.

In the construction sector, the opportunity for using machine learning technology is immense. Many construction projects usually have many issues that need to be handled carefully and correctly. In such works, many changes may happen without prior information. Manually addressing such challenges and improvements can be tedious and mistakes occur. Through implementing machine learning systems and allowing project management technology to take care of whole programmes, routine processes can be largely automated and human error potential can also be effectively mitigated. Machine learning systems can support project managers, bosses, and practically anyone interested in such tasks when used in construction projects. Using such technology undoubtedly saves time. Such software solutions can also be used to track job performance, making it easy to

Table 13.1 Benefits of machine learning in construction

Code	Meaning
MLB1	Risk assessment
MLB2	Productivity booster
MLB3	Safety maintenance on site
MLB4	Documentation and record keeping
MLB5	Inventory stock taking
MLB6	Generative design
MLB7	Prevents cost overruns
MLB8	Project planning
MLB9	Labour shortages amelioration
MLB10	Offsite construction

detect mistakes or risk factors. Managers and supervisors can get feedback on such problems and swift corrective action can be taken and this will help to avoid complications or cost increases. Although machine learning is an emerging science and computer technology, its current construction uses are not negligible. The forms in which most building companies now incorporate machine learning software to their ventures are described later (Rajagopal & Tetrick, 2018; Aniruddh, 2019).

Table 13.1 as well as Figure 13.1 represent various sustainable benefits of machine learning for construction projects. It includes factors such as risk assessment, productivity booster, prevention of cost overruns, and inventory stock taking among others.

Risk assessment

In building, before a construction project is approved by the authority and designers take action, a risk assessment must be undertaken. This applies to nearly any residential or commercial project. It was done manually previously, which was time-consuming. Software tools are also used, but machine learning technological implementation has rendered the risk assessment more reliable and quicker. This is because machine learning software tools will collect and review huge amounts of data to provide a detailed and accurate risk assessment for such initiatives. It can be used to assess and measure construction-related risk factors such as soil type and building height feasibility (Alexander, 2019; Ellis, 2019; Rao, 2019).

Productivity booster

If building management technology enabled by machine learning is used in a construction project, it improves the productivity level. Such software solutions can be used in worksites to track or control everyday activities. They can be used to track activities such as brickwork, masonry, plumbing construction, electrification, flooring, and roofing. The system generates an alarm and preventive action can be taken if anything goes wrong. It further eliminates the risk of accidents and also helps to save time. Workers in the company will concentrate on their respective roles and ultimately become more successful (Rajagopal & Tetrick, 2018; Alexander, 2019; Aniruddh, 2019; Ellis, 2019; Rao, 2019).

Safety maintenance on site

Many risk factors occur in most construction sites, irrespective of position or form. These include electrocution hazards, flame failure, short circuits, coping with heavy equipment, and even the worst-case scenario, death. While it is not uncommon to install surveillance cameras or fire alarms, the use of machine learning software tools brings the safety level to a new height. The approach can be used to check for risk factors and it is possible to identify and document any deterioration immediately. It aims to mitigate the level of damage or disaster at such places (Rajagopal & Tetrick, 2018; Rao, 2019).

Documentation and record-keeping

Immense amounts of data must be processed and viewed regularly in any construction project. Mistakes may happen when this is done manually, and this contributes to many problems. Using accounting software is an alternative; however, mistakes can be made if it is done by humans. When using machine learning software saving and extracting information is trouble-free. These tools can also be used to gather specific data from a massive server and review or modify existing data very quickly. In fact, anything relevant to the programs is electronically recorded and saved – which in the future can be helpful in solving any problem (Aniruddh, 2019; Ellis, 2019).

Inventory stock taking

Large quantities of supplies must be processed close by or at the project site in most construction projects. These include bricks, concrete, colours, sheets of iron, and blocks. It is very important to keep control of the stock. Having workers to do this leaves room for mistakes. However, it makes things easier by introducing machine learning software to manage task stocks (Alexander, 2019; Aniruddh, 2019; Ellis, 2019; Rao, 2019).

Use of Autonomous machinery and robotics

The installation of specialized equipment and robots powered by machine learning software is becoming more prevalent in large-scale and commercial construction ventures than in the past. It helps to eliminate the risk of accidents or casualties and also improves the success rate of the project. It is feasible to use machine learning-driven robotics to build parts of buildings at elevated heights. This also frees human workers to concentrate on doing more complex work such as drainage and setting up electrification (Rajagopal & Tetrick, 2018; Alexander, 2019; Aniruddh, 2019; Ellis, 2019; Rao, 2019).

Generative design

Generative design is a method of form-finding that can imitate the adaptive approach to the layout in nature. Computer scientists have found ways to support the process of building construction. This usually begins by clearly defining the aims of the project and then investigating various possible permutations of a solution to find the best alternative. Awareness modelling is a 3D model-based system that offers technical insights into architecture, technology, and development to better prepare, develop, create, and maintain buildings and

infrastructure. The 3D designs should take into account the structural, technological, mechanical, electrical, and plumbing (MEP) plans and the series of operations of the relevant teams in order to schedule and lay out the development
of a house. The goal is to ensure that there is no overlap between the various
models from the sub-teams.

Prevents cost overruns

Despite hiring the most qualified project managers, many megaprojects go over
budget. For programmes, synthetic connected networks are used to forecast
cost overruns when considering some factors such as task duration, the form of
contract, and project managers' level of competence. Predictive models utilize
historical data such as expected commencement and end dates to visualize practical schedules for future projects. Machine learning allows employees access to
realistic training material directly and enables them to improve their skills and
knowledge easily. It reduces the time spent on task engaging new resources. As a
consequence, the completion of the plan is improved upon (Rajagopal & Tetrick,
2018; Aniruddh, 2019; Ellis, 2019).

Project planning

A machine learning start-up was introduced in the year 2018 with the hope that
its robotics and artificial intelligence would have the key to solving belated tasks
in construction as well as over-budgeting. The company employs robotics to take
3D images of construction sites autonomously and then feeds the information
into a deep connected network which sorts the differences between the various
sub-projects. When items are off course, before they become major issues, the
management team will move in to help with small issues. Potential algorithms
will use a machine learning technique known as reinforcement learning, which
enables test-based and error-based testing for algorithms. Based on similar ventures, it can compare infinite variations and alternatives. As it optimizes in the
best way possible, and corrects itself over time, it assists with proper project planning (Alexander, 2019; Aniruddh, 2019).

Labour shortages amelioration

Labour shortages and the importance to improve under-productivity of the sector
are convincing constructing companies to engage in machine learning and data
science. According to a 2017 McKinsey survey, construction firms can elevate
productivity by up to 50% by realistic data collection. Building businesses are
continuing to use machine learning to better plan research and equipment delivery through employment. A robot that continuously monitors task advancement
and the position of staff and equipment helps project managers to tell immediately that job sites have adequate employees and equipment to make sure that the

project is completed on time. This will reduce menace in construction especially when additional work is introduced. Researchers predict machine learning technologies will render building robotics smarter and more autonomous (Alexander & Ellis, 2019).

Off-site construction

Firms are beginning to depend on factories situated offsite where there is manufacturing of autonomous robots that are then installed on site by human workers. Structures such as walls can be more effectively completed by robots than using human labour, allowing humans to focus on simpler and more detailed work such as ventilation, HVAC, and electrical systems when the building is completed (Rajagopal & Tetrick, 2018; Alexander, 2019; Ellis, 2019).

Machine learning and big data in construction

Machine learning systems were subjected to infinite amounts of data to benefit from and evolve every day at a period where large amounts of data are being generated daily. Every site where construction work is ongoing becomes a possible source for machine learning information. Data generated through mobile device-recorded images, safety sensors, drone clips, building information modelling (BIM), and others have become an information database. It provides an opportunity for practitioners and customers in the construction industry to evaluate and take advantage of the lessons produced from the information by utilizing machine learning and machine learning systems (Aniruddh, 2019; Ellis, 2019; Rao, 2019).

Post-Construction infrastructure management

Managers can use machine learning even after a project has been built. Urban management software, or BIM, holds data about the building's design. Machine learning can be used to track challenges and even provide approaches for problem avoidance. Machine learning can also be influential in facilities planning to expand an asset's complete lifecycle past design and construction. Generally, important information in facilities leadership often has gaps. As a consequence, on-site maintenance and upgrades are difficult to manage quickly and cost-effectively. Through efficiently gathering and using information and data details, machine learning may help streamline the process. It can do this with remarkable precision by classifying documents and data such as work order and analysing related circumstances in real-time. It removes away from people these boring, time-consuming administrative duties and helps us to focus on the real issue at hand. In fact, when machine learning is implemented into a BIM system in operations and maintenance, the best way to conduct maintenance and repairs can also be defined by visualising when and where problems occur (Rajagopal & Tetrick, 2018; Alexander, 2019; Rao, 2019).

Project management

Building project management has a host of factors to address, with each having the potential to stall a project for years. Without breaking a sweat, machine learning can perform all these activities. Potential infrastructure systems can oversee whole facilities including design costs, buildability, and architectural integrity for various technical approaches to large-scale commercial projects, individual residences, and ventures outside the country (Alexander, 2019; Aniruddh, 2019; Rao, 2019).

Quality control

The quality control system for a manufacturer or company as well as its potential inhabitants is repetitive yet critical. Neural networks can help with this method, the cornerstone of machine learning itself. Neural networks could be used to associate drone-collected photos with existing models and compare different design irregularities. Before they occur, builders or managers will have the ability to detect any flaws or likely threats to a house, reducing time and cost (Rajagopal & Tetrick, 2018; Ellis, 2019).

Business model design

An artificially intelligent system will probably understand the customers better than you do. Today machine learning is used in multiple industries to fully understand the needs of consumers, building personalized product interactions. The intelligent construction future will also be made from this algorithm. The new age of development and implementation will be to understand the needs of a consumer based on data and also to consider how to satisfy them. Intelligent architecture can anticipate consumer preferences, sometimes even in real-time, adjusting the business model to the market A designer utilizing machine learning can determine which product mix is most appealing to a client (Alexander, 2019; Aniruddh, 2019; Rao, 2019).

Limitations to machine learning

Even with the growth experienced over the years that has in turn directed the attention towards machine learning, it is quite clear that machine learning is still subjected to some limitations regarding its usage and application. For illustration:

- The dependence of approaches that work within the concept of machine learning in relying on granting permission to access the numerous data available, in conjunction with the creation of accurate information can be resource-intensive as well as time-consuming.
- It could be quite difficult to create a system that fully understands and relates to a specified problem. In cases of expertise failure, humans fall back on

applying their contextual understanding (common sense) of the situation and will often take actions, which while they may not be the best, are rarely the cause of any major damage. Currently, the system of machine learning does not possess or encode this behaviour meaning that when they fail, they may fail in operations in a serious manner. Simply put, the machines work in relation to the program installed into them.

- Humans possess the ability to transfer ideas from a problem source to another with much ease. This remains a major challenge for these computers even with the latest upgrade in the machine learning techniques available now with little or no human aid.
- Imputation of many constraints in the actual sense of reality into natural laws expressed in physics, mathematical, and other related laws such as logic is difficult in machine programming. It is not the most correct to consider removing these constraints when considered from the angle of machine learning.
- Comprehending the intent and wants of humans is highly complex, more than what a designer might input in a machine. It requires a sophisticated understanding of taking into consideration every possible attribute and relating such intelligence into a formidable aspect of learning. The available methods are limited towards a pointed domain of the machine according to its operational systems.

Conclusion

Machine learning may sound like a far-off idea of architecture, decades away from becoming a fact. The truth is the technology's future is close. In the construction industry, machine learning has been steadily gaining attention. While it may seem like a non-human solution that is highly technical, it can actually make things more human. Machine learning allows people to do their real jobs more efficiently instead of taking people out of the equation. Machine learning can help the industry to flourish, improving things for employees, contracting companies, and end-users on a daily basis. Machines can basically understand and interpret outcomes on their own. Machine learning uses algorithm-based software instead of human programming that helps it to make predictions based on its data analysis. The technology coupled with artificial intelligence will help make worksites more effective and, in the end, save money.

References

Alexander, D. (2019). *Smart construction: 7 ways AI will change construction.* Retrieved from: https://interestingengineering.com/smart-construction-7-ways-ai-will-change construction

Aniruddha, R. (2019). *AI and machine learning: What's in it for the construction industry?* Retrieved from: https://geniebelt.com/blog/ai-and-machine-learning-whats-in-it-for-the-construction-industry

D'Amico, B., Myers, R. J., Sykes, J., Voss, E., Cousins-Jenvey, B., Fawcett, W. & Pomponi, F. (2018). Machine learning for sustainable structures: A call for data. *Structures*, 19, 1–4.

Ellis, G. (2019). *How machine learning is making construction more human*. Retrieved from: https://blog.plangrid.com/2019/02/machine-learning-in-construction/

McAfee, A., Brynjolfsson, E., Dvavenport, T. H., Patil, D. & Barton, D. (2012). Big data: The management revolution. *Harvard Business Review*, 90(10), 61–67.

Novotny, R. (2019). The importance of sustainable construction. eSUB constrution software. Retrieved from: https://esub.com/blog/the-importance-of-sustainable-construction

Rajagopal, A. & Tetrick, C. (2018). *The rise of AI and machine learning in construction*. Retrieved from: https://www.autodesk.com/autodesk-university/article/Rise-AI-and-Machine-Learning-Construction-2018

Valpeters, M., Kireev, I. & Ivanov, N. (2018). Application of machine learning methods in big data analytics at management of contracts in the construction industry. *MATEC Web of Conferences*, 170, 01106.

14 Nanotechnology for sustainable construction

Introduction

The construction industry globally is assumed to be the highest single facilitator of some of the problems encountered in the environment due to the effect it has on survival. Buildings accounted for worldwide are calculated to be liable for about 50% of the energy consumed by the consumer, and more than 50% of emissions experienced globally, as well as consuming between 30% and 40% of the electric energy consumed globally. Pollution of the environment, felling of trees, soil erosion, ozone degradation, and human health risks are most affected by faulty designs, erections, and overseeing of buildings that increase the impacts on the environment and its constituents. Obviously, activities in construction constitute a crucial role in our current environmental plight.

There has been a high level of popularity of nanotechnology among researchers and the industrial sector in the last few decades. The number of nanomaterial products entering the market is also increasing rapidly and in the years to come the phenomenon will be even more pronounced. Nanotechnology, according to the Oxford Dictionary, is the engineering division that deals with dimensions and tolerances of less than 100 nanometres, in particular, the processing of individual atoms and molecules. This consists mainly of the extraction by one molecule or one atom of product separation, aggregation, and deformation (ANNIC, 2019). In our modern world, nanotechnology is one of the most effective breakthrough drivers and offers a ground-breaking alternative to long-standing and real-world solutions. Sustainable development cut across three different social systems: society, environment, and economy. Sustainable construction is the response of the construction industry to sustainable development. According to the Environmental Protection Agency (EPA), sustainable construction is said to be the practice of constructing structures and implementing practices that are responsible environmentally, the management of resources in relation to efficiency, and the total number of years that the building will serve from the site down to the design, development, service, repair, reconstruction, and deconstruction (Environmental Protection Agency (EPA), 2018). A model was designed

DOI: 10.1201/9781003179849-14

by Kibert (2018) for sustainable construction and compared with the traditional concern in construction in terms of:

- Cost
- Performance
- Quality

According to Kibert (2016), some principles were set for sustainable construction as follows:

1 Pursuing quality in creating the built environment (Quality),
2 Maximising reuse of the resources (Reuse),
3 Minimising consumption of resources (Conserve),
4 Using renewable or recyclable resources (Renew/Recycle),
5 Protecting the natural environment (Protect Nature), and
6 Creating a healthy, non-toxic environment (Non-Toxics).

The construction industry has much to gain from nanotechnology, among other industries, to ensure sustainability in construction. Nanotechnology is one of the latest construction industry practices to enhance the worldwide application of sustainable construction.

One often cited definition is that of Brundtland's report in 1987 as cited by Kono (2014) where sustainability was defined as development that meets the needs of the present generations without compromising the ability of the future generations to meet their own needs. He further defined sustainability as follows:

i An all-embracing concept covering all the technical understanding of everything concerned that elucidates a healthy, dynamic, and desirable equilibrium between human and natural systems;
ii A dynamic system consisting of most exclusive practices, reasoning with the aim to safeguarding the various components and sects that makes up the ecosystem to boost the economical aspect of any industry coupled with the opportunities in it, and therefore creating a better life for whoever is going to use it; and
iii Discussing a future in which anyone can live happily and also partake.

Drafting conclusions from the definitions above, sustainability entails:

- Community and individual human well-being,
- Economic or financial considerations, and
- Environmental protection and stewardship.

These are the three recognized factors of sustainability. This explains much in improving the economic and social impact in the quality of life and confining impacts on the environment to the carrying capacity of nature. The construction industry has much to gain from nanotechnology, among other industries, to ensure sustainability in construction. Nanotechnology is one of the latest construction industry practices to improve the worldwide application of sustainable construction.

Nanotechnology

Nanotechnology is a technique enabling us to create materials with modified or completely new properties. Many scholars consider nanotechnology to be a modern science and technology, but it has been deduced from further studies that nanotechnology is not a new found science, nor is it newfound technology, but rather an advancement and extension of science and technology that has been under progress for decades. It is a very complex field of engineering, the application of which encompasses many fields. It is estimated that the total market size for nanotechnology goods will rise to $3,000 billion in 2020 (Roco, 2011).

The concept of nanotechnology first came into play in a speech made by Richard Feynman in his seminar entitled 'There is plenty of space at the edge' in 1959 in which he outlined the capacity to synthesise material by the direct manipulation of atoms. The term 'nanotechnology', however, didn't resurface until Norio Taniguchi used it in 1974 (Taniguchi, 1974). Stirred by Feynman's theories, in his book titled *Engines of Creation: The Coming Age of Nanotechnology*, Drexler employed the term 'nanotechnology' to suggest the concept of a nanoscale self-assembling particle that could create a replica of itself and other atomic-contained objects (Drexler & Minsky, 1990). According to Wikipedia, nanoparticles are microscopic particles the size of which are measured in nanometres (nm) and which have always been present in the world. It is defined as a particle of less than 200 nm in at least one dimension. The type of semi-conductor materials between a conductor and an insulator, e.g. silicon may also be known as quantum dots if they are negligible enough (typically below 10nm) to cause a surge in energy levels. The benefit of this when energized is that, for example, a UV light will emit different colours from the same substance in different sizes.

Roco (2011) reported five key indicators in respect to the ways of developing nanotechnology everywhere. These indicators are listed below:

1 The number of people working in the sector,
2 The number of scientific publications in journals and other publications,
3 The number of filed patents,
4 The number of final products depleted in the market, and
5 The financing for research and development.

Rico provides an estimation of the value obtained in 2015 in the table and an extrapolation to the value of 2020. Given the global crisis in 2008 and 2009, this sector's rapid growth is impressive. According to the report by Roco, in just 20 years, the labour force prices and the products that are finally obtained on the market will increase about 100 times. In terms of research, the number of articles quadrupled between 2000 and 2008, while patents and funding for research and development increased tenfold (Roco, 2011).

Nanotechnology for sustainable construction

Nanotechnology generally has also proved its usefulness as it is used in construction in making quality, advanced, and well-structured materials for construction activities. Bringing the use of nanotechnology into construction can be applied in many materials to include concrete, steel, timber, coating, paint, and glass. The application of nanotechnology to sustainable construction comes with many benefits to be derived from its usage and application. The three areas of sustainability will thus benefit from it, but its application to the construction sector would be classified as the environmental area of sustainability.

Importance of nanotechnology in sustainable construction

According to Ezel (2014), the potential importance of the adoption and the usage of nanotechnology in elevating the construction industry are as follows:

1   The employment of nanoparticles, nanofibers and carbon nanotubes to enhance the durability and strength of construction components and materials as this will reduce pollution to the environment;
2   Affordable manufacturing of steel that does not corrode with exposure to corroding agents;
3   Manufacturing of heat materials that are insulated to give better performance;
4   Ability to minimize pollution and energy consumption by coating it with coats that are thin and can clean, capable of changing colour; and
5   Making materials that are of the nanosize with sensors that can repair themselves to increase dependability.

Environmental

The environmental area refers to practices that do not jeopardize the today at the expense of tomorrow's generational use. The environmental area of sustainable construction considers materials used to mitigate emissions of gases and reduce the impact on the environment. Nanotechnology can be effectively used in producing materials that can improve the environment without causing harm to the environment. The following are areas in which nanotechnology can be used to improve building materials.

Structural building materials

Structural materials are crucial in building for it to achieve the goal it is meant to achieve in terms of longevity and thus the sustainable nature of the structure. Elvin (2007) stated that taking into consideration the strength of the materials and their weight ratio, it is crucial that materials that have a stronger and lighter capacity will definitely have a greater load carrying capability per unit material. Nanotechnology increases the ability to improve structural materials in two ways:

i   By adding reinforcement to the already existing materials, such as wood, concrete, steel, glass, and light, with the fusion of nanoparticles to enhance its properties; and
ii By providing new materials like carbon nanotubes (CNTs) that are structured to cater greatly with the goal of being technically and economically feasible.

However, concrete is undergoing drastic enhancements with the advent of nanotechnology as it is the largest production (annually) among other materials classified a structural material.

Bhuvaneshwari, Sasmal and Iyer (2011) indicate the application of nanotechnology to develop several building materials among which are the following:

Nano cement

Changing the particles present in cement to that of nanosize will increase its surface area. For example, combining nanotubes and reactive nanosize silica particles can help to achieve this feat. Using nanocement with the correct concrete mix is way better than using nanocarbon tubes or carbon fibres in concrete. Oke, Aigbavboa and Semenya (2017) also suggested that the properties of concrete can be adjusted in large numbers by infusing nano particles such as nano silica and nano clays to give surface functionality. During the past decades, the performance of concrete has been essentially enhanced by applying various types of miniaturised scale and nanoparticles in other to strengthen and enhance its general durability.

Nanotechnology for concrete

In contrast to steel, the construction industry uses concrete as a fundamental material. It is also estimated that 3.3 billion tons of the concrete binder that is cement are manufactured worldwide. As of 2009, there is around 8% growth, and the overall bond production continues to expand, powered by developing economies such as those of China and India. Thanks to its nano-sized structure consisting of cement hydrates, aggregates, and additives, concrete has become an excellent candidate for nanotechnological engineering (Péro, 2004).

The element properties and concrete hydrate on the nanoscale have a strong influence on the concrete itself as a macro-material (Beaudoin, Raki & Alizadeh, 2009). This means that its properties are strongly influenced by the hydrating gel of calcium silicate, a nanosized composite material and it's significantly altered by a multi-scale network of capillary pores and microcracks (Zhu, Bartos & Porro, 2004). In fact, concrete is an omnipresent man-made building material (Sanchez & Sobolev, 2010), which is nano-structured (at similar scales) and consists of many aging phases over time. Concrete properties exist in and processes of degradation occur across several length scales (nano to micro to macro) where each scale's properties derive from those of the next smaller scale (Sanchez & Sobolev, 2010).

Nano steel

In reducing the uneven surface of steel, adding copper nanoparticles will work well by reducing the unevenness accordingly. It also limits the enhancement in steel stress and thus reduces fatigue cracking. In vanadium and molybdenum, nanoparticles have tremendously improved the fracture problems emanating from high strength bolts as experienced frequently prior to the use of nanoparticles. Researchers have reported that stronger and better steel cables can be produced by refining the cementite phase of steel into a nanosize. After the second industrial revolution in the late 19th and early 20th centuries, steel has been readily available and has since played a major role in the construction industry as it has been used in various ways and forms. Despite the fact that steel is always under high stress, fatigue is an important element that can contribute to steel's structural failure when subjected to cyclical loadings, such as bridges or towers and the like. Subsequently, by including copper nanoparticles, the steel fatigue will be reduced. At the juncture when copper nanoparticles are combined with the steel, the surface of the completed steel becomes smoother and even more planed. Hence, this surface evenness may reduce stress, and subsequently the steel may have less pressure and breakdown due to fatigue.

Nano glass

According to Oke, Aigbavboa and Semenya (2017), the application of nanotechnology may bring about an increase in energy efficiency by conserving building heat energy through a reduction of heat loss. By using glass that is manufactured from nanotechnology, there is a reduction in the energy that is consumed by the ventilation, and as well as the air conditioning system. This is made possible as solar heat is prevented from entering due to heat controller which functions in absorbing the heat. Nanotechnology has been in use for a long time in the application and production of coatings created to offer features for improved glass performance. Glass can be manufactured to have sterilising, self-cleaning and anti-fouling features by formulating the presence of TiO_2 in nano form as a form of cover over it.

Item design in the glass itself has concentrated over the past few years on meeting an increasing functionality scope. This entails manufacturing glass with improved execution in terms of warmth protection, solar control (to reduce heat loss and maintain ventilation), safety and security, fire impermeability, noise reduction, reflection, self-cleaning, scratching, and visual appearance (Andersen, 2011). The integration of nanomaterials into glass matrices enables windows to be developed with the ability to control how light and heat move through building walls. These technologies can lead to a plus in energy efficiency, for example, saving heat energy by reducing heat loss, using nano-produced low emissivity glass and reducing energy consumption.

Application of nanotechnology to other materials

Nano lighting

Through nanotechnology advances highly efficient LEDs can be produced that will have considerable savings on the quantities of the electricity generated and used. This is made possible by the inclusion of heat sinks that helps a hot fan to stay cool whenever it is in operation. OLED (Organic LED) technology is applied in various sectors already; however, it is unfortunate that is only used in small or minute quantities. Since we are in an era where technology is progressing at a swifter rate, the OLED light applied for several uses will save an unfathomable amount of energy and this will bring a newer perspective to the lighting system, not just in construction but in any related industry. OLEDs are put together to function when there is a passage of current to the layered part of the nanoscale films. It can be utilized to show at the surface of which it set or even can be made to be transparent to visibility. And from the surface, it can serve as the source of light that is also applied in some windows to replicate the kind of light received during the day even when it is night time (Soutter, 2012).

Nanotimber

According to Oke, Aigbavboa and Semenya (2017), using nanotechnology in a forest can improve the products for the industry bringing unimagined growth and advancement opportunities for bio-based products. Nanotechnology will bring about a one of a kind up-and-coming era of wood-based items that have hyper-performance and predominant serviceability when used in harsh environments. Expanding the accentuation on wood as a building material could have noteworthy ramifications for worldwide vitality prerequisites and worldwide carbon dioxide discharges through nanotechnology improvement which can be made to offer clearly defined economic, environmental, and sustainability benefits with respect to its usages in the construction industry. Other potential advantages will be enhanced conventional wood- and bio-composite items like those now utilized as a part of construction, however, with altogether enhanced performance or economy.

Purifying Water

With the use of nanotechnology, the quality of water to be consumed daily has improved a great deal. There are photo-catalysts that can serve as oxidizing agents to neutralise organic pollutants into that which is safe and not dangerous. These photo-catalysts are as follows:

- Titanium dioxide (TiO_2)
- Zinc oxide (ZnO)
- Carbon nanotubes (CNTs)
- Silver nanoparticles
- Zeolites
- Nanoparticles of zero valent iron (ZVI)
- Tungsten oxide

TiO_2 is the most preferred material because of its high photo-stability and high photo-conductivity; moreover, it is inexpensive and not harmful. Silver nanoparticles have also been said to possess antimicrobial properties that help in killing microbes during the process of water treatment in the secondary stage of treatment. Also, many other polymeric nanoparticles are used to make sure that wastewater treatment is more effective to provide cleaner water for consumption.

Another technology, newly known as nanofiltration, can be applied in treating water, especially in construction industries for use in construction work. Molybdenum disulphide (MoS_2) is a non-porous membrane introduced for energy efficient desalination of water which filters five times more than the conventional ones (Ray, 2019). Nanotechnology can also improve the field of electronics and computer memory and that will surely be hugely beneficial in constructing sustainable buildings, hence green building.

Conclusion

The numerous uses, impacts, and applications of nanotechnology in sustainable construction are exemplified in materials science and engineering towards enhancing the innovative capacity in achieving sustainable construction. Materials of various calibres can be modelled with desired properties that would modify their related functionality, which will help in achieving practice that is directed towards sustenance. Moreover, the economic implications of mixing the forms of nanotechnology to crude construction materials is that cheap and accessible infrastructural developments will be achieved through the reduction of material productions costs. In view of this, the applications of nanotechnology should be considered by construction stakeholders, including contractors, clients, consultants as well as agencies concerned with monitoring, managing, and regulating construction activities in the quest for the continuous growth and advancement of the construction industry.

References

Andersen, M. M. (2011). Silent innovation: Corporate strategizing in early nanotechnology evolution. *Journal of Technological Trans formation*, 36(6), 680–696.

Applied Nanotechnology and Nanoscience International Conference (ANNIC). (2019). *5th edition of Nanotechnology &Nanoscience Conference in Paris*. Retrieved from: http://nanowerk.com

Beaudoin, J. J., Raki, L. & Alizadeh, R. A. (2009). 29 Si MAS NMR study of modified C–S–H nanostructures. *Cement and Concrete Composites*, 31(8), 585–590. https://doi.org/10.1016/j.cemconcomp.2008.11.004

Bhuvaneshwari, B., Sasmal, S. & Iyer, N. R. (2011). Nanoscience to nanotechnology for civil engineering: Proof of concepts. In *Proceedings of the 4th WSEAS International Conference on Recent Researches in Geography, Geology, Energy, Environment and Biomedicine (GEMESED'11)*, 14–17 July, Corfu, 230–235.

Drexler, K. E. & Minsky, M. (1990). *Engines of creation*. London: Fourth Estate.

Elvin, G. (2007). *Nanotechnology for green building green technology forum, 2007*. Retrieved from: http://esonn.fr/esonn.fr/esnn2010/xlectures/mangematin/Nano_Green_Building55ex.pdf

Ezel, M. (2014). Nanotechnology innovations for the sustainable buildings of the future. *International Journal of Architectural and Environmental Engineering*, 8(8), 886–896.

Kibert, C. J. (2016). *Sustainable construction: Green building design and delivery*, 4th edn. Wiley. Texas, USA.

Kibert, C. J. (2018). *Establishing principles and a model for sustainable construction*. Retrieved from: https://www.irbnetde.com

Kono, N. (2014). Brundtland commission (world commission on environment and development). In: Michalos, A. C. (Eds), *Encyclopaedia of quality of life and well-being research*. Dordrecht: Springer. https://doi.org/10.1007/978-94-007-0753-5_441

Oke, A., Aigbavboa, C. & Semenya, K. (2017). Review of the application of nanotechnology for sustainable construction materials. In *Proceedings of the2nd International Conference on Mechanics, Materials and Structural Engineering (ICMMSE 2017)*, Atlantis Press, 102, 364–369. http://ujcontent.uj.ac.za8080/10210/388219

Péro, H. (2004). *Nanotechnology to new production systems: The EU perspective, nanotechnology in construction*, Special edn. Edition No. 292. Scotland: Springer.

Roco, M. C. (2011). The long view of nanotechnology development: The national nanotechnology initiative at 10 years. *Journal of Nanoparticle Research*, 13, 427–445.

Sanchez, F. & Sobolev, K. (2010). Nanotechnology in concrete – A review. *Construction Building Materials*, 24(11), 2060–2071.

Soutter, W. (2012). *Nanotechnology in green construction*. Retrieved from: https://azorano.com

Taniguchi, N. (1974). On the basic concept of nanotechnology. In *Proceedings of the International Conference on Production Engineering, Part II. Japan Society of Precision Engineering*, Tokyo, 18–23.

Zhu, W., Bartos, P. J. M. & Porro, A. (2004). Application of nanotechnology in construction summary of a state-of-the-art report. *Material Structure*, 37(273), 649–658.

15 Robotics for sustainable construction

Introduction

Over the past few decades, there were many multi-level structural building systems. The systems and material production for the structures consume energy and there were emissions of carbon from the onset of the construction processes. This led to a global trend of the elimination of non-renewable and waste during the manufacturing and construction process using these materials. There is a difference in the level of energy needed to manufacture them, and these energies are passed down to the entire service duration of the building. It is necessary to assess the life cycle and sustainability of the structure. Sustainable construction can be illustrated as the process of building structures using systems that are eco-friendly and energy-intensive throughout the entire whole durational service of the building, from inception to design to construction to occupancy, maintenance, rehabilitation, and finally, demolition. It can also be defined as the types of construction that are designed to lower health and environmental impacts generated by the construction process or structures or built up the climate. The key concept of sustainability is the conservation of preserving everything concerned with nature and the environment, the use of materials that are free from harm, the re-use of resources, the minimization of waste, and getting knowledge of whole life cycle of the building.

This study concerns the involvement of working towards sustainable construction in relation to robotics engaged in construction processes and achieving sustainable construction through robotics (Mills-Tettey, Dias & Nanayakkara, 2005). Sustainability simply refers to the present necessities today without neglecting the needs of the future. The awareness of sustainability emerged in the 1970s amidst the oil crisis; it was at the time the need for and urgency of energy saving was paramount; and the need is to make use of energy and explore alternatives to prevailing energy sources. Three sustainable construction criteria were resource depletion, environmental pollution, and a safe environment. Six principles were laid down for sustainable construction as follows:

1 Minimize resource consumption,
2 Opt for resources that are reusable,

DOI: 10.1201/9781003179849-15

3 Opt for the use of renewable or recyclable resources,
4 Preserve the habitat,
5 Ensure a healthy and non-toxic environment, and
6 Work tirelessly on creating a built environment of quality.

There is considerable evidence that sustainable buildings generate economic benefits to building proprietors, managers, and tenants. Buildings working towards sustainability usually have lower or reduced running costs for power, water, refurbishment or repair, modification (reconfiguration of space due to changing needs and conditions), and other cost of operating (Dzioubinski & Chipman, 1999).

Reduction and re-usage

Sustainable building is not just about ensuring that our projects use resources efficiently. This further entails pulling all the strings concerned with the environment and impacts generated by viewing opinions in considering products and all the concepts employed in getting the job done. This means having a vivid comprehension of the project not only with regard to the bottom line and the gain to the business but also with regard to the way it corresponds to things in the environment, society, and the planet, considering all possibilities (Nair & Potter, 2011). Hence the reason why sustainable construction is so important is because it affects everything present or in the environment, and not just where the building is being erected.

Construction firm's role in sustainable construction

Like any business, the organization needs to show a profit before anything else. That being said, investing in green technology can have significant advantages when it comes to building. Leadership in Energy and Environmental Design (LEED) architecture is the current industry benchmark for green building. LEED controlled the sector by setting the benchmark for sustainable building practices since its introduction in 1994 as it encourages a green construction system that exceeds beyond the minimum or stipulated building codes to ensure that new buildings are not only functional today, but are also in line with sustainability in the future, energy efficient on all fronts and not only the built industry, and built from responsible sources. Builders who are actively partakers in the construction process in the latest sustainable technology will recover these costs over time as an outcome of higher energy output in the wake of lower building operating costs. Construction firms today need to consider an increasing and growing concern for sustainable building (Bordass, 2003).

Government role in sustainable construction

Through creating the right avenues for companies that opt to invest in a sustainable way, regulators play a significant role in sustainable building. Organizations doing this should be encouraged and praised. People should be given the opportunity to 'vote with their wallet' when it comes to engaging organizations that comply with the ideals of sustainability in projects. Simultaneously, the government may impose regulations by legislating policies that require companies to build in a sustainable manner. Some states are presently making available tax incentives and exemptions for constructing companies using sustainable practices. Across Nevada, construction materials can be bought with a tax exemption for structures that are certified as sustainable by LEED. In Cincinnati, Ohio, newly built buildings that satisfy minimum LEED requirements, are offered property tax exemptions. Such exemptions permit construction firms to invest in buildings directed more towards sustainable infrastructure and make available a net benefit for all stakeholders, including the public and sectors associated with the environment.

Green construction and sustainability

Green building takes a step further than sustainability and relates how in the construction process we can mitigate our environmental impact and, in some situations, how we can offset the effect we have. Green construction establishes a high standard in every form of sustainability, finding green building materials and methods (principles) that are carbon neutral or require renewable energy source present now and later in the future (Dzioubinski & Chipman, 1999).

New innovations

The construction via LEED is not the only way green construction can take place. In recent construction advances have been made that can help mitigate the quantity of materials used and thus improve construction sustainability. In making it easier for enterprises to improve project sustainability, these and other developments should be considered.

Prefabrication

For building, prefabrication refers to being environmentally friendly. The method for assembling units within a plant is prefabrication instead of doing it on site. So, a factory that is involved in prefabrication can build nearly any necessary element and transport it to the place where it is assembled or requested. The highlighted advantage is that the units are built or manufactured in a warehouse, so that no waste is thrown out. Alternatively, in another unit, it can be recycled and made use again. Prefabrication also assists the building to be more long-lasting, so less material has to be reused.

Building information modelling

Designs made by the use of building information modelling (BIM) software has put the industry towards revolutionising the way constructions are planned, estimated, and constructed. BIM is far becoming more reliable and acceptable as it assists with the required material and time calculations that is big advantage to the professionals implementing its use. This can reduce the amount of purchasing excess building materials and increase the building's overall design as well reducing unnecessary stress.

Four pillars of sustainable construction

Sustainable construction integrates four socially oriented systems that are the environment, society, human, and economy (otherwise known as planet, people, partnership and peace, and profit) with a framework aimed at realising a sustainable construction.

Environment sustainability

In a truly sustainable world, the ecosystem can preserve species, biodiversity, and general productivity over an extended period of time. Ideally, choices must have an equilibrium effect on the biological systems and try to foster significant growth. Environment disruptions should be eliminated; noise should be reduced to the minimum degree permissible. The environmental impacts or consequences of all decisions should be discussed. There are a number of items that are directly connected to environmental sustainability, namely proper management, preservation, and protection of natural resources in order to propagate better productivity. The objectives of environmental sustainability are to keep assets intact.

Social sustainability

It strives to maintain wealth by spending and developing programmes that form the structure of our society. The theory represents a wider view of the word in relation to nations, cultures, and globalization. This means protecting generations to come and realising what we do will have an effect on someone else and the environment. Social sustainability emphasises developing and preserving social values with principles such as unity, mutual respect, fairness, and the relevance of human relationships. Statutes, knowledge, and collective thoughts on justice and freedoms can be encouraged and supported. The theory of sustainability is based on the notion that an action or plan encourages the development of society; in particular, generations to come should have the same or higher living standard benefits as the present generation. The definition of social sustainability often covers other topics such as civil rights, environmental law, and public policies.

Human sustainability

It addresses the question of how to improve quality of life, how human population is affecting the habitat, as well as addressing issues on nutrition, knowledge, and skills, access to services, leadership, and health and education systems. Human sustainability aims to balance the general well-being of citizens and the available natural resources. It eliminates the consumption of toxics materials that end up contaminating our environment and avoids uncivilized behaviour, disputes, and wars that can lead to famine, death and poor health of the citizens, and other aftermaths of war. It also pursues equality and fairness for all regardless of ethnicity, race, or gender. The issues mentioned above are what we come across on construction projects year in year out. Human sustainability expects the organization to view themselves as part of the community and the society at large, treating every employee as important, and ensuring that their manufacturing process does not endanger society.

Economic sustainability

The objective is to keep assets intact as it centres on improving the quality of life. It entails creating economic value from projects or choices you make. It implies that decisions taken are the most efficient, effective, and fiscally feasible while taking into account other facets of sustainability. Having a high and steady degree of economic development is part of the major goals of sustainability; nevertheless, sustainability is much more important than economic development. Modern economics incorporates natural resources, ecosystems, and social connections and questions the capital maxim that continuous growth is good. Solid financial aid for schools, training programmes, and research and development are also significant aspects of sustainability.

The practice of sustainable construction

Practice in this concept relates to the ways sustainable construction is being carried out on project sites, from the onset of the design stage throughout the entire cycle of the project. They are discussed below:

1   The design of the building must take into cognizance the impact the structure will have on the environment; for instance, well-insulated buildings require minimal energy to heat or cool spaces in them. The design must take into consideration conserving water, use of renewable materials, natural aeration, and illumination in the building as these will minimize running costs throughout the building life as well as contributing to sustainable construction.
2   The use of materials that have a negligible effect on the environment should be promoted during the time that the construction is ongoing. The manufacturing process of construction materials generally consumes more energy and with the aim to lower energy consumption, the use of materials that can be changed from one form to another to be used again, reused or converted to other materials should be promoted. In addition, materials from naturally occurring elements should be adopted and salvaging from

other sites materials that are unused (this will also contribute to waste reduction). For instance, palm kernel in large quantities can be used for concrete in place of chippings and fly ash can be added to concrete to improve on its elasticity and minimize the quantity of cement. Also, the use of locally sourced materials should be encouraged to reduce or minimise the transportation of materials across different states, thereby keeping our environment safe.

3 Powering equipment on-site using wind, hydropower or water will minimise energy consumption. Exhaust fumes can be trapped from pipes and treated with a catalyst to become water and reused.

4 Waste reduction systems can be employed, whereby waste is properly stored as they are being generated and reused. Carefully deconstruct or decommission components before demolition, so as to save the components for reuse. Reuse or give out the materials in good and usable conditions, preferably within the shortest time. Recycle whatever cannot be reused. Many components or materials can be generated through this method for reuse (Bordass, 2003).

The future of robotics in construction

Technology takes over the building step by step and changes the way a project is performed. The need for manual labour is projected to decrease significantly in the near future as a large amount of work will be carried out automatically. Robots can provide great assistance in tasks involving precision and acceleration. In fact, we can easily overcome on-site problems like human limited strength or physical weariness.

The meeting Point between humans and robots

To achieve success in using robotics in building works, there is a clear need to incorporate the robot and human additions in such a way as to complement one another. It can be challenging to find the right meeting point. To maximize efficiency, we need to put together the top-performing characteristics both hands. A thorough categorization of each side's different strengths and weaknesses will help to get as close as possible to a harmonious human-robotic cooperation.

Robotics engagement in construction processes

Robotics is a multidisciplinary engineering and science field encompassing mechanical, electronic, software engineering and others. It is concerned with the creation, production, and use of robots that use systems of computing for their programming, control, data processing, and sensory inputs. Robots are useful for several purposes but most importantly, they are used in volatile situations such as detecting and coping with places where humans might find it difficult to survive, e.g. under water, in space and at high temperatures and the cleaning and checking of toxic chemicals and radiation.

Today, robotics is a fast-expanding field as innovations continue, studying designs, and creating new robots to meet a range of practical purposes; be it domestic, industrial, or military. Industrial robots are common now and use to execute tasks more efficiently, accurately, and timely. They are used for jobs that are too filthy, toxic, and boring for humans. Examples of jobs that robots can be engaged on include seeking survivors in dangerous ruins, discovering mines and shipwrecks, to mention a few. The use of robots on site will not eradicate humans working on a project; rather they complement each other, and the highest performance of each will be exploited in order to maximize efficiency. Advantages of using robots on site are the following:

1   **Minimize mistakes**: Robots can give much needed guarantees of accuracy which when considered, minimizes human errors out of the anything related to construction process, thereby minimizing mistakes and cost.
2   **Reduce cost for the construction process:** The use of robots requires a heavy capital outlay. However, in due time, the overall cost of the project can be reduced through minimising project delays and with better time efficiency required for the completion of a task that also limits expenses.
3   **Protect the workforce:** Involving robots in the construction activities will ensure the benefit by making sure that they stay in good shape health-wise because robots will do the heavy manual work with humans to supervise them. Workplace safety will be improved owing to the fact that some of the unsafe tasks on site will be automated.
4   **Improve the industry's practice:** The construction industry is a slow and monotonous field. This poses a challenge of promoting its profile. Robots being introduced to the construction process will make construction appealing to the younger generation to pursue a career in connected to new technology. The arrival of youths in the industry will provide solutions to the labour shortage problem that has been crippling the industry for years.
5   **Meet deadlines:** The automation of the construction process will permit managers to plan with effectiveness, with an increase in precision, and project deadlines will be met (Bordass, 2003).

Although the use of robots on construction sites is encouraging, the high cost of acquiring and training staff on the use of robots coupled with the fear of employees losing their jobs to a machine are threats to the adoption of robotics for construction.

Disadvantages of using robots in construction

1   **Fewer jobs within construction:** The introduction of robotics in the construction process will provide exciting opportunities for people who are interested in technologies that haven't been seen before, but correspondingly will

put an end to a significant number of on-site jobs. Of course, robotic technology cannot fully replace the human involvement, but it can certainly limit it to a considerable degree.

2 **Cost of acquiring the robotic machines:** It was found before that during the management of the construction process, robots can play a beneficial role in planning. But the cost of purchasing the equipment, however, continues to be extremely high. Therefore, there is a possible chance of acquiring the best technology with fewer companies having the financial capacity to purchase it. With these, many small or medium-scale construction companies may eventually go out of business and that will increase the ever-increasing unemployment.

3 **Workforce training:** Buying is one thing; acquiring the skills for operating the machines is another thing. For companies that are ready to train and involve their staffs on using this cutting-edge technology, a significant amount of time and money will be needed. This said company would aim to recover the money spent and could therefore lead to reducing the quality of work expected as they hope to recuperate money spent and also make profits from it.

Examples of robots that can be used on construction sites

Masonry robots

SAM 100 (semi-automated masonry system) masonry robots are designed to set up to 350 bricks per hour in regular or soldier brick pattern. That is a lot faster than most human bricklayers. It is programmed to carry bricks to the point of use, lift a brick, apply mortar, and set on site. It is designed to place the bricks accurately and clean up the excess mortar. There are already robots with the ability to perform masonry work on the market. Their use is not yet widespread, but on numerous construction sites around the world they are already offering great support. The pace and precision of their motions are truly unique.

Doxel AI

Doxel artificial intelligence (AI) tracks progress on site with real-time workable data, using stand-alone drones and rovers fitted with high-definition cameras and LiDAR (light detection and ranging) sensors to scan, capture and monitor a project site with precision each day. Then the AI compares the scan with the project BIM models, 3D drawings and estimates to examine the quality of work done and progress made each day. It will also pinpoint if there are discrepancies between work done on site and the project drawings. The AI uses advanced algorithms to recognise and document errors performed on site as it can identify building components based on their form, size, and location. By grouping and measuring quantities Doxel AI can determine the quantity of work done daily and notify if a project is lagging behind.

Automated track loader (ATL)

ATL is created to excavate on small project site, with specially built LiDAR that can withstand vibrations to see where it is going, spot movement and measure excavated materials. Enhanced GPS with on-site base stations and satellites combined are used to geo-fence the site and move the track loader around the site with precision. The ATL system is installed in a cargo carrier that could be connected to an existing track loader instead of buying a piece of new equipment. The machine also has a collision detection system that prevents the machine from colliding with workers or other equipment around the site. It is designed to operate with a remote control system, with equipment operators input so the machine can operate at the same speed or faster as a human operator. The design of autonomous trucks is already doing a commendable job. In the coming years, they will be a serious game-changer. Their usage allows for improved on-site production and lower operating costs. Improved safety is also a major advantage of autonomous vehicles for people working on site.

TyBot

A reinforcement tying robot is designed to work with only one labourer to supervise. The TyBot is set up on the project, and it goes from one bar intersection to the other, tying the bars with binding wire. It is designed to work tirelessly until the whole reinforcement is properly tied. It is brought in after the reinforcement has been put in place.

Street-laying machine

The Tiger stone is a brick road paver designed to pave 3,229 square feet of brick road in a day with just two human operators. The bricks have to be fed to the robot by hand in the desired pattern from a hopper to a pusher slot and then the bricks are laid in one continuous layer. Street-laying machines are an example of a robotic machinery variant that is already used in building. The device is still supplied with bricks by hand, but the rest of the job is fully automatic. This encourages a better, quicker and higher output.

Building drones

The design was proposed back in 2012 by the company Gramazio Kohler Architects and was soon part of the on-site daily reality. They showed, with the aid of the robotic expert Raffaello D'Andrea, that drones could create solid architectural constructions simply by working together. Such an innovation would completely change the way we do on-site work. Improving the on-site output, the completion time of construction project can be reduced drastically. These are like drones on construction sites. Flight-assembled architecture is a drone that is designed as an architectural installation using quad-rotor helicopters that communicate with each other in order to lift and assemble components to create a structure.

3D printing robots

Conventionally construction projects involve several steps which range from when the planning is done, to development to on-site construction and this has been the same over the years. The creation of 3D printing robots is the change construction industry needs; the robots have the ability to build in the toughest conditions and require fewer materials to build more resistance and unique structures. They prefabricate components on site, thereby eliminating transportation of large components or materials and within a lesser overhead cost and completion time. If the client does not like how the prefab turns out, it is easy to disassemble and reassemble.

Achieving sustainable construction through robotics

Robotics can assist companies to incorporate sustainability practices into their daily activities. There are robots that conserve energy, minimise wastes, construct green structures, make use of less materials because they work with precision, record daily activities, measure on the site, compare data with the project model to ensure conformity, and eliminate errors. Some are used for recycling purposes because they can work in dirty and dark places and they can pick reusable parts from discarded materials. Their roles in achieving sustainable construction are and cannot be limited to:

Lean construction practices

Lean construction is centred on the elimination of waste. Engaging robotics on a project site can help lower the amount of waste generated on the site of work, because robots have the ability to work with accuracy and precision. They make use of standardised components that will positively impact the environment and the economy at the same time.

Reduction of climate-related risks

The use of advanced technology to build and upgrade infrastructure will assist in reducing climate-related risks. A climate-stress absorption and adaptation technology are already integrated into the construction industry. Automation and robotics possess a significant footprint which can improve construction without affecting the climate. The robotics are designed and programmed to recognise, absorb, adapt and evolve under stress, in a way that contributes to the sustainability of the environment.

Reduction of energy consumption

Robots can work tirelessly and efficiently in the cold and dark places, thereby making it possible to conserve energy. Some robots can perform basic tasks like cleaning, and mowing lawns. There are some being used on construction sites as a means of transportation so as to reduce energy consumption and emission. There is a correlation between energy efficiency, renewable energy sources and climate change; robots can provide all these benefits, thereby reducing contamination and emissions through the monitoring of the release of toxic greenhouse gasses and refining the manufacturing process.

Use of the passive solar system

Passive solar machines are technologies that convert sunlight into usable heat, make air-movement for aeration and using a little energy from other sources. It is aimed at minimising the way people use non-renewable resources and energy and improving the life cycle of the project. The non-renewable resources on sites are land, water, materials and energy; the preservation of these will be vital for a sustainable future.

Conclusion

Resources (energy and material) are limited in a sustainable economic system. Robotics can help, but for them to play a useful role, they will need to be tailored to suit needs. Industrial robotics and automation are currently making a quantifiable increment in human workers' productivity, which may in fact have a destabilizing effect on the economy. Nevertheless, robots also contribute to development in a qualitative way. In theory, sustainable production management does not prohibit robots from being used. The objectives of sustainable construction are designing for human adaptation, which means providing a safe and comforting structure for humans with social amenities that are not harmful. Energy conservation is another key objective of sustainable construction. Renewable materials must be chosen for construction. Insulating a structure will help to prevent heat loss and the heat trapped in the kitchen and bathrooms can be mechanically used in other spaces of the building. Materials conservation is also a key objective; lean construction must be actively practised on all projects sites. At the design stage waste minimization of waste must be taken into consideration, recovering construction waste to reuse on site, reusing and recycling products and the last objective will be to store all construction waste properly so as not to cause harm in the near future.

References

Bordass, G. (2003). *The economics benefits of sustainable design.* Retrieved from: http://www1.eere.energy.gov.femp/pdfs/buscase_section2.pdf

Dzioubinski, O. & Chipman, R. (1999). Trends in consumption and production: household energy consumption. *United Nations Discussion Paper*: ST/ESA/1999/DP.6.

Mills-Tettey, G. A., Dias, M. B. & Nanayakkara, T. (2005). Robotics, education, and sustainable development. In *Proceedings of the 2005 IEEE International Conference on Robotics and Automation*, 18–22 April, Barcelona, 4248–4253.

Nair, C. & Potter, R. (2011). *Consumptionomics: Asia's role in reshaping capitalism and saving the planet.* Hoboken: Wiley.

16 Virtual reality for sustainable construction

Introduction

Virtual reality (VR) is a computer-generated simulation of a three-dimensional environment, in which the user is able to both view and manipulate the contents of that environment (Goulding et al., 2012). The benefits of VR can be seen in areas such as architecture, education, and field engineering processes that range from design to final construction. VR brings to attention a particular aspect observed to be viewed later by bringing such into view now even before construction in order to make the unreal look more like reality. VR creates an environment where computer-generated information (CGI) is superimposed onto the user's view of a real-world scene (Chi, Kang & Wang, 2013). The benefits of VR can never be over-emphasized as it gives a clear view of the desired end product even at the initial conception (planning) stage of the project. VR is engaged in the construction practice by creating a three-dimensional model that aids in bringing more reality to the theoretical part of the project by bringing an idea to reality as near to reality as it can. Park et al. (2013) explained that virtual interpretation is more exclusively experienced than the total theoretical state it was before.

Development of VR

Visualization simply explains the process of giving a design another and better perspective to data obtained or designed in a project by offering a different view for clarity, transformation/alteration, and corrections of known or unknown faults. Virtual technology is not an alien technology. It has been in existence for a long time. The popularity just depends on the number of users that use the virtual development in whatever aspect of their service. In fact, the actual base of VR dated back to 1962 with the ill-fated 'sensorama' of Morton Heiligs and the introduction of tele-operation displays by involving the use of head-mounted television systems and closed circuits.

The application of VR in communication, design modelling, and visualization has received a great deal of interest and development recently. One of the

DOI: 10.1201/9781003179849-16

major reasons why VR has engendered enormous support is due to the fact that it offers benefits to a considerable wide range of users such as scientific visualisation through much equipment; operations carried out in hazardous, unplanned, and remote areas; and particularly in planning, putting the plans into place, and designing ideas in a built environment before they are established or taken into consideration.

User's experience can be achieved by using a range of contrastive VR hardware solutions such as haptic devices, sound systems, tracking systems, and stereoscopic displays. VR entails two main types basically explained in immersion and interface in a synthetic environment, and the second being in immersive and non-immersive VR systems:

- Immersive VR is an application in relation to quasi-physical experience. The experience contains the closest contact that occurs between the person who is operating the system and the virtual environment to be related to. This is a type of VR that can be obtained by engaging data gloves and multimedia head-mounted display (HMD) devices.
- Non-immersive VR adopts a screen interface that enables users to have a feel of the modelled environment with the aid of additional devices such as the eyeglasses. VR of this type is also known as desktop VR or screen-based VR that encompasses 'division' and 'superscape' software for better operations and additional output.

VR systems could also be classified into HMDs (head-mounted display) and cave automatic virtual environments (CAVEs). HMDs include Gear VR, HTC Vive, Oculus Rift, and Google Cardboard that provides a cost-effective VR alternative by inputting the head-mounted gears that are attached to the operator.

VR and planning smart cities

In planning the best type of cities and the environment as a whole, there is every need to put in place plans that are concurrently related to giving a better platform for a developed and interactions among the components that makes up the entire environment. The creating of an environment that is sustainable is vital as well as making provisions through the aid of the technologies present. The process or act of providing effective visualization tools to evaluate and predict the environmental and social results of creating a city is the key to achieving a sustainable and resilient urban design using the limited available land. The traditional method of approach in visualization is to display information or data within a 2D framework concept. The 2D layers are not that explanatory and shortened in knowledge as the data or information encoded in it only covers a specific item of work when compared to what a 3D model offers. Moreover, the possible use of 3D VR system has been researched to overcome the elements that are inadequate in the use of 2D systems.

By using VR, planners have the potential to identify the extent by which a new concept introduced is affecting the subjected part. They can map and at the same time provide behavioural responses, ethnical perspectives, and generally the social concerns about the planned concept. VR gives a platform to an environment that can be accessed and related to with the information at hand by enhancing and making further modifications in facilitating the visual aspect and making evaluations on the new design schemes.

Impact of virtual reality on Green building construction

The use of VR in the construction industry has improved the nature of the construction process to an international standard. Some of the benefits that can be derived from its utilization include clients' experiencing swifter construction processes and architectural firms developing the way they design constructions for their clients with the objective of meeting the ever-increasing demands.

Benefits of utilizing virtual reality within the construction industry

VR simply functions by changing the *view* of your brain by making it relate in a way that presents a three-dimensional environment that is different from the environment you are actually in. Angles of various projections are imbedded into it to create this three-dimensional effect and therefore it gives a precise view and an in-depth perception of all the different elements being displayed, seemingly bringing reality in front of you. Gaming and the entertainment industry have predominantly employed this, so can the construction industry. Here are some of the highlighted benefits of using VR technology in construction:

Viability and feasibility appraisal

From the inception of a project, the viability and feasibility of the project are first considered among other factors as they are crucial to whatever plans and decisions that would be made with regard to the project. Using VR at the design stage allows architects and designers along with other professionals to picture the viability and feasibility of the project by considering factors that are concerned with the environment. With this, better accuracy is obtained than using a combination of human judgment and scale models, which presently is the only option available for judging the viability of a project. VR leaves practically no room for error and can fully imitate the finished concept as it would be due to its nature of driving home the real appearance of the project.

Modify the design at any stage of the construction

VR also assists the construction professionals that are involved in bringing the project to reality to make confident decisions about the progress of work at any stage of the design. These decisions aid in altering the design, plans, or materials if need be before it is built. VR provides transparency and allows clients to feed back without wasting time, resources, and money on things that may later be changed when the problem could have been spotted earlier.

Safety and construction practicalities

In preventing any safety issues that might occur along the line of construction now or later, VR enables construction project managers to pick up on any potential hurdles along the way that could be easily averted or avoided, whatever the case may be. For example, in a situation where there is a small space for the construction workers to carry out the part of the project needed within that space, or large elements may be tricky to manoeuvre through areas of the new building, all of these can be solved by making alterations to the design, or working out a different or alternative method of doing things.

Improved project planning

It is no longer news that through its cost and time minimization, VR has improved the quality and possibly quantity of projects to be erected within the shortest period of time. Both the clients and the operator in charge would have assessed the best fit for the project thereby moving them to a point of total or near total certainty.

Enhance collaboration

Implementing VR brings along with it other forms that are working towards an identified goal between the professionals involved and the clients as they get to relate better to what and where their services are required. For the clients, VR puts them in a situation whereby they can relate to what they are actually paying for in terms of employing the services of the construction professionals in having knowledge of the project rather than the typical drawings or plans that most cannot comprehend. For many clients, it is difficult to interpret the lines and coding drawn on plans, but VR brings the specific vision of a client to life by showing an exclusive view of practically everything, as it also provides enough platform for sharing ideas, asking questions, and making corrections to faults before the actual plan is put in place.

Factors mitigating against the wide use of virtual reality

Even though the benefits of engaging VR in construction are well spelt out, it could not help some obstacles that negate its benefits. They are listed as follows.

Public perception

Some clients doubt its *seriousness* since VR originated from the video game industry. The clients feel sceptical about its adoption. This naivety is not only applicable to the clients alone; some construction companies are not considering its adoption either. It has been proven over the years that changing people's perspective about a particular issue can be a herculean task. Nevertheless, the opinions and minds of the client could be addressed towards its benefits once they see and experience its operational usefulness, proving once and for all the need to employ it in construction.

Technical know-how

Most firms do not have the technical capacity to handle the operations involved in the systems designed in VR. The cost of engaging professionals to cater for this particular aspect of the job is expensive, especially for a small-scale firm. Therefore, this has made managers of these firms to stick with what they practise in terms of using traditional ways of construction, hence walking away from an industry that is on the path to technology.

Non-participation of young professionals in the construction industry

As mentioned earlier, younger generations that are technology savvy will prefer to engage in construction practices that involve the use of VR and therefore leaving the *older generation* in their wake. And this might make some of them to stop operation as they might not be able to operate this software or the cost of hiring professionals to operate it for them might be something that they cannot afford.

Conclusion

The VR brings to a reality that which was assumed to be a mirage. It incorporates the necessity to have a complete and relatable perspective of a construction model with the basic details available. VR does more than just bringing a perceived reality to the fingertips; it connects the parts together in order to arrive at an acceptable inference on the type of project embarked upon. VR in construction has been helping to get the best of the inputs (material and labour) integrated into the construction system by providing a platform that can be easily comprehended among those that are involved in construction.

References

Chi, H. L., Kang, S. C. & Wang, X. (2013). Research trends and opportunities of virtual reality applications in architecture, engineering, and construction. *Automation in Construction*, 33, 116–122.

Goulding, J., Nadim, W., Petridis, P. & Alshawi, M. (2012). Construction industry offsite production: A virtual reality interactive training environment prototype. *Advanced Engineering Informatics*, 26(1), 103–116. http://dx.doi.org/10.1080/21573727.2013.805688

Park, C. S., Lee, D. Y., Kwon, O. S. & Wang, X. (2013). A framework for proactive construction defect management using BIM, virtual reality and ontology-based data collection template. *Automation Construction*, 33, 61–71.

Index

Note: *Italicized* and **bold** pages refer to figures and tables respectively.

For Product Safety Concerns and Information please contact our EU
representative GPSR@taylorandfrancis.com
Taylor & Francis Verlag GmbH, Kaufingerstraße 24, 80331 München, Germany